Requirements Engineering for
Safety-Critical Systems

RIVER PUBLISHERS SERIES IN SOFTWARE ENGINEERING

The "River Publishers Series in Software Engineering" is a series of comprehensive academic and professional books which focus on the theory and applications of Computer Science in general, and more specifically Programming Languages, Software Development and Software Engineering.

Books published in the series include research monographs, edited volumes, handbooks and textbooks. The books provide professionals, researchers, educators, and advanced students in the field with an invaluable insight into the latest research and developments.

Topics covered in the series include, but are by no means restricted to the following:

- Software Engineering
- Software Development
- Programming Languages
- Computer Science
- Automation Engineering
- Research Informatics
- Information Modelling
- Software Maintenance

For a list of other books in this series, visit www.riverpublishers.com

Requirements Engineering for Safety-Critical Systems

Editors

Luiz Eduardo G. Martins

Federal University of São Paulo, Brazil

Tony Gorschek

Blekinge Institute of Technology, Sweden

River Publishers

Published, sold and distributed by:
River Publishers
Alsbjergvej 10
9260 Gistrup
Denmark

www.riverpublishers.com

ISBN: 978-87-7022-427-7 (Hardback)
     978-87-7022-426-0 (Ebook)

©2021 River Publishers

Contents

Preface

Requirements in safety-critical systems play a central role. Not only they specify how the system can safely interact with its environment but they play a fundamental role in the verification and validation of such system's safety. They are also fundamental to certification processes, which cannot be carried out without complete and precise requirements.

Though there exist several books on requirements engineering, this book focuses on aspects that are specific to safety-critical contexts. It addresses complementary issues including the interface of requirements engineering with hazard and safety analysis, certification based on safety cases, and automated testing. It also covers practices in several key application domains such as aerospace and medical equipment, as well as the dependency between security and safety.

Because its material is rooted into practice and its focus is on safety-critical aspects, I recommend this book to both practitioners and researchers who would like to gain a good understanding of both the state of the art and advanced practices in the context of safety-critical systems.

Lionel Briand
Professor, Canada Research Chair, IEEE and ACM fellow.
University of Ottawa and University of Luxembourg

Acknowledgments

The KKS Research Profile SERT (Software Engineering ReThought) at SERL, Sweden, Blekinge Institute of Technology, was a partner in research contributing to this book. For more information see rethought.se

List of Figures

List of Tables

List of Abbreviations

AC	Advisory circular
AD	Acitvity diagram
ALARP	As low as reasonably practical
ARP	Aerospace recommended practice
ASCAD	Adelard safety case development
BML	Behavior model language
BMP	Behavior modeling processes
BT	Behavior tree
CAE	Claim-argument-evidence
CASE	Computer-aided software engineering
ConOps	Concept of operations
CPU	Central processing unit
CSW	Components of software
DPAL	Data process assurance level
DQR	Data quality requirements
DRM	Design reference missions
DSML	Domain-specific modelling language
DSN	Deep space network
EASA	European aviation safety agency
EDAC	Error detection and correction
EMC	Electromagnetic compatibility
EMI	Electromagnetic interference
ESA	European space agency
FAA	Federal aviation administration
FDA	Food and drug administration
F-E-F	Faults, errors, and failures
FTA	Fault-tree analysis
GLCD	Graphic liquid crystal display
GPS	Global positioning system
GSN	Goal structuring notation
HA	Hazard analysis

HAZOP	HAZard and OPerability study
HLA	High level architecture
HLR	High-level requirements
ICSW	Item configuration of software
IEC	International electrotechnical commission
IEEE	Institute of electrical and electronics engineers
IIP	Insulin infusion pump system
IMA	Integrated modular avionics
IP	Internet protocol
ISO	International organization for standardization
IV&V	Independent verification and validation
JPL	Jet propulsion lab
LCD	Liquid crystal display
LLR	Low-level requirements
MBD	Model-based development
MBT	Model-based testing
MCO	Mars climate orbiter
MDD	model-driven development
MHRA	Medicine and health regulatory authority
MPA	Main process area
MPL	Mars polar lander
MTTF	Mean time to failure
MTTR	Mean time to repair
NASA	National aeronautics and space administration
NEAR	Near earth asteroid rendezvous
NFR	Non-functional requirements
NFR	Non-functional requirements
OAO	Orbiting astronomical observatory
PDI	Parameter data item
POFOD	Probability of failure on demand
QFD	Quality function deployment
RAD	Requirement activity diagram
RBT	Requirement based testing
RE	Requirements engineering
RF	Radio frequency
ROCOF	Rate of occurrence of failures
RQ	Research question
RTCA	Radio technical commission for aeronautics
SACM	Structured assurance case metamodel

SAE	Society of automotive engineers
SC	Safety cases
SCD	Safety case development
SCS	Safety-critical systems
SEU	Single-event upset
SIL	Safety integrity level
SLR	Systematic literature review
SPA	Sub-process area
SPI	Software process improvement
SRSt	Software requirements standards
STAMP	Systems-theoretic accident model and processes
STAR	Self-testing-and-repairing
STPA	Systems-theoretic process analysis
SW	Software
SysML	Systems modeling language
TBT	Test behavior tree
TC	Test case
TDD	Test-driven development
TMR	Triple modular redundancy
UML	Unified modeling language
UNIFESP	Federal university of são paulo
Uni-REPM	Unified requirements engineering process maturity model
Uni-REPM SCS	Uni-REPM for safety-critical systems
USW	Unit of software
XP	eXtreme programming

1

Introduction

Safety-critical systems (SCS) have gone mainstream. In essence, most people today are dependent on, use, and are enabled by, SCS. Traditionally SCS were seen as separate and special, demanding rigorous conceptualization, engineering, and quality control, as the implications of failure were severe resulting in harm, loss, or significant detriment. However, the evolutionand saturation of use, of all manners of software intensive products and services (SIPS) has changed the landscape. Today, most systems can be seen not as optional or addons, but rather as utilities. Mobile phones, networks, and services are one such example that might be obvious todaybut were considered a side-note 25 years ago. The same is true for communication apps, productivity systems, and access to online services. This smudging of boundaries, where a consumer smartwatch in essence complements, or even supplements, medical equipment is omnipresent as technology and access pushes the traditional categorization.

This evolution has implications, which we will discuss shortly, but more importantly it has potential and benefits. The ability for invention and innovation is accelerating and far outweighs any detrimental effects. For example, linking a smartwatch to an app that tracks health stats, and through cloud-based big data analytics using narrow AI and machine learning predicts a heart attack. The rigor of such "consumer" SCS is less, however, the level of access to a wide range of users has enormous potential for good. Functionality traditionally associated with medical domain grade systems in this case has moved to the public domain. Further, the interaction between benign parts networked can result in critical functional combinations. The interaction and functional networking of multiple independent systems, such as third-generation collision avoidance, combined with route planning and mobile positioning to offer safer and more economical routes to a driver. These are two small examples of when the boundary of critical and consumer based systems blur.

1

There are two main implications to this. First, many companies and the SIPS they supply are not considered to be in the category of SCS. These companies do not have the tradition of SCS engineering, nor are regulated as such. Second, many companies that are considered to supply critical systems will be subject to competition (alternatives), but more importantly functional networking of offerings, where their products will be one part of a chain, will impact their offering and ability for quality assurance. This might present the instinct to introduce more regulation and control, which might be a part of the future, however, a better and more mutually beneficial approach would be to improve the overall engineering in all cases. Companies and organizations with critical systems expertise can both learn and share experience and knowledge.

This book presents an important step in the evolution, but also sharing, of critical systems engineering knowledge, and starts at the logical and important knowledge area of requirements engineering - figuring out what to include in a software intensive product or service, and to what level of quality. There are countless examples of the importance of requirements engineering as a knowledge area that is inherent in all of the parts of the engineering chain - from the inception, planning, design, realization, and verification and validation and evolution of a system. Thus we will not motivate it further here, but rather focus on the development of the area in the context of this book.

In the fast-moving pace of (traditionally non-SCS) SIPS we see that good-enough requirements engineering and the use of iterative continuous engineering and delivery mechanisms utilized as a part of "agile" or "lean" practices and principles are outpacing more plan-driven approaches. The motto "fail fast, fail often" is a good summary of trying out solutions instead of focusing on catching the requirements "correctly" and validating them. This way of thinking is rather contrary to more rigid engineering approaches where formalism is dependent on "getting it right" and validating correctness before committing to development. One could argue for both approaches, and try to force one over the other - however, both are a reality. Both have benefits. Good-enough approaches focus on end results pace and offer the realization that resources and time-to-market are critical for success. Rigid approaches focus on avoiding the potential detrimental effects of making mistakes, which in certain applications take precedence.

However, since critical and non-critical SIPS get more and more indistinguishable, there is a need to evolve the respective engineering practices. This book offers an important part of this continuous journey.

But it is done in a responsible and useful way. It takes the deep and significant tradition of critical systems engineering and elaborates towards today's organizations developing SIPS by making it more available outside of the traditional fields of SCS. The book also looks at opportunities to introduce new and improved practices and "good-enough" perspectives to the SCS community by bringing ideas and concepts from requirements engineering to the table in a manner compliant with standards, regulations, and principles. The second part of the book offers important application examples in various domains of application that anchors the book beyond theory and generic standards. The book is also significantly based on recent research into the field of requirements engineering for SCS.

We would recommend this book to three categories of professionals and students. If you are active in the SCS category of SIPS the book offers you insights into requirements engineering as an enabler to improve practices and principles already in placeand offers a bridge between rigid and more continuous engineering. If you are working in a non-SCS application area, you can learn a lot from SCS engineering, and most likely the future will necessitate this knowledge. If you are in between, a life-long student of new ways of combining good ideas from different outlooks, you might find synthesis and inspiration to fill your engineering toolbox.

2

The Role of the Safety and Hazard Analysis

Jéssyka Vilela

Universidade Federal de Pernambuco, Brazil
E-mail: jffv@cin.ufpe.br

Abstract

The complexity introduced using software to control hardware components leads to accidents and safety incidents caused by requirements problems. This chapter explores the role and influence of safety and hazard analysis in the safety requirements definition. The relationship between safety/hazard analysis and requirements engineering is explored by defining safety and hazard analysis methods, which results in safety requirements that should be specified early in the software development process to develop safer systems.

Keywords: Safety Analysis, Hazard Analysis, Safety Requirements Definition, Requirements Engineering, Dependability.

2.1 Introduction

Companies have produced even more highly evolved computing systems used in people's daily lives, requiring high reliance [1][3]. For instance, to provide a set of cash dispensers, control a satellite constellation, an airplane, a nuclear plant, o a radiation therapy device, or to maintain a sensitive database's confidentiality. In this type of computing s, called Safety-Critical Systems (SCS), failures (events that occur when the delivered service deviates from its specification) could cause accidents resulting in damage to the environment, financial losses, injury to people, or even the loss of lives [1][2].

Accordingly, these safety-critical systems should support the dependability property, which corresponds to the ability to deliver services that can justifiably be trusted [3][4] as well as analyze and specify safety requirements correctly [2]. Dependability is usually decomposed into six attributes [3]:

- Availability: the ratio of time a system is ready for use.
- Reliability: the probability of absence of failure in software operation for a determined time in a defined environment.
- Safety: the absence of severe consequences to users and the environment due to system usage.
- Confidentiality: the lack of unauthorized disclosure of information.
- Integrity: The level at which a system or component hampers unauthorized access to, or modification of, computer programs or data.
- Maintainability: the ability of systems to be subjected to modifications and repairs.

The intended purpose for the system will direct the emphasis considered in each attribute [3]. This chapter emphasizes the *safety* attribute aiming to increase the ability to avoid failures that could be catastrophic to users or the system's environment [3]. In the following sections, we establish the foundations of safety engineering by describing the safety-related concepts involved in hazard analysis and safety requirements specification.

2.2 Foundations of Safety Engineering

A system may fail because it does not comply with the specification or because it did not adequately describe its function [1][3]. A causal chain - Faults, Errors, and Failures (F-E-F) - threatens the dependability of a system in the sense that the chain completion leads the system to a state that reports incorrect service or outage [4]. This causal chain, described in the next section, is of fundamental importance for applying safety and hazard analysis techniques discussed in Section 2.3.

2.2.1 The Threats: Faults, Errors, and Failures

In the F-E-F chain, illustrated in Figure 2.1, a fault is the cause of an error, a state of the system that may lead to failure. In this causal view, an error is seen as an intermediate stage between fault and failure. However, it is essential to note that a fault does not necessarily lead to an error, which does not necessarily lead to failure.

A *fault* is the adjudged or hypothesized cause of an error. A fault is active when it produces an error; otherwise, it is dormant [3]. We can

Figure 2.1   The causal chain of threats (adapted from [4]).

cite a physical defect that occurs within hardware components and design faults/implementation mistakes as examples of faults. Avizienis et al. [3] provide a very rich and precise taxonomy of sources of faults, presented in Figure 2.2.

The erroneous behavior characterizes error states for components, connectors, and services due to fault occurrences [4]. Sometimes a component raises an error that does not reach the system boundaries, meaning it does not cause a failure. This occurs when the service delivered by the component is not in the system interface. However, such a component may offer its service to another internal component, leading to error propagation between components [4]. At some point, this error will lead to the occurrence of a failure. Finally, *failure* is a deviation from the system's specification resulting in incorrect behavior regarding expected correct operation.

For safety-critical systems, fault, error, and failure concepts are complemented with safety-related concepts such as hazards [4] described in the next section.

2.2.2 Safety Concepts

The specification of a safety-critical system involves the analysis and documentation of safety-related concepts. In a previous work [10], we present Safe-RE, a safety requirements metamodel elaborated based on industry safety standards, whose aim is to support the specification of safety-related concepts in the RE process. Safe-RE consists of entities and relationships

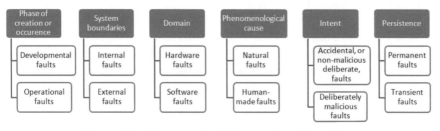

Figure 2.2   Elementary fault classes (adapted from [3]).

that represent concepts common to different safety standards from other domains. The Safe-RE metamodel contains information that must be recorded regardless of the safety standard being followed [10].

Safety analysis aims to identify, analyze, and specify the safety-related concepts and demonstrate that the developed system is safe. The main concepts [10] are defined below and illustrated considering an Insulin Infusion Pump system (IIP) [8] [9]:

- **Accident**: it is an undesired and unplanned (but not necessarily unexpected) event that results in (at least) a specified level of loss. Some accidents can occur due to the use of the IIP, such as *user receives incorrect treatment, user infection, electrical shock, and environmental pollution.*
- **Safety Incident**: It is an event that involves no loss (or only minor loss) but with the potential for the loss under different circumstances. On the other hand, **harm** can occur to **assets** of the system or the Mission.
- **Hazard**: it is a system state that might, under certain environmental or operational conditions (**context**), lead to different consequences: **accident**, **safety incident,** or cause **harm**. It is associated with Functional Requirements and has a **severity** degree and **likelihood**. Several hazards can occur in the IIP, such as Overdose: the user receives more insulin than required; Underdose: the user gets less insulin than needed; Excessive thermal energy generation by the pump; Excessive sound frequencies generated by the pump; and Excessive pump vibration.
- **Risk**: it is a combination of consequence (hazard severity) and **probability** of the **hazard**, i.e., risk = hazard probability x hazard severity. The probability of sensor failure in IIP that may result in an insulin overdose is Occasional. Therefore, the **risk** of an insulin overdose is probably medium to low.
- **Safety Functional Requirement**: It should be implemented to avoid the occurrence of hazards or to mitigate them. For example, considering the *insulin overdose* **hazard** and one of its causes is the *free flow*. Some **safety functional requirements** to mitigate them can be the *operation of the pump within a temperature between 25 °C and 37 °C;* and, *manage of reservoir volume.*

2.3 A Method for Safety and Hazard Analysis

Considering that Safety-Critical Systems (SCS) are usually submitted to safety certification processes, in their development, an in-depth Hazard Analysis (HA) is required to ensure that the system is safe and the hazards of the system were appropriately mitigated.

Safety engineers have used several techniques in safety analysis during the last decades. Safety analysis aims to identify safety needs that ensure or mitigate that system failures do not cause injury, death, or environmental damage. We can consider four main steps to be conducted during the safety analysis, illustrated in Figure 2.3 and explained in the following sections.

2.3.1 Step 1: Hazards Identification

The *hazards identification* step consists of performing the hazard analysis in which experienced stakeholders, working with engineers, safety experts, and professional safety consultants, identify hazards according to their experience and from an application domain analysis. Many techniques could be used at this stage as recommended by the international standard IEC-60300-3-1. Some criteria may be considered for technique selection since there is no best technique for all systems. The technique selection depends on the needs of the company and the system. Some criteria that could be used are presented by Bernardi et al. (2003) [4]:

- **Applicability in the Life Cycle**: The *early stages* techniques are used in the requirement analysis or when the design process starts; The

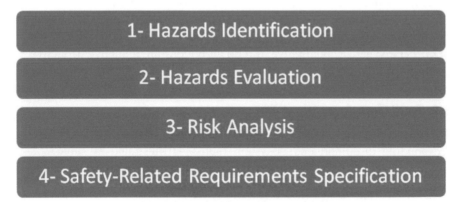

Figure 2.3 Main steps of safety analysis.

late stages techniques are used when the detailed design specification becomes available. Finally, techniques can be used across different stages: initial models are constructed during the requirement analysis and, then, refined to a more detailed level as data become available to make decisions and trade-offs.

- **Aim of the Analysis**: It can be either *qualitative* (used to the identification of the component failure modes, their consequences at a system level, and the cause-effect paths, as well as the determination of possible repair/recovery strategies) or *quantitative* (used to define numerical reference data to be used as input parameters of reliability/availability models).
- **Bottom-Up/Top-Down:** *Bottom-up* methods mainly deal with the effects of single faults, and the starting point of the analysis is the identification of the faults at the component level. *Top-down* methods, on the other hand, enable analyzing the effects of multiple faults.
- **Cause-Effect Relationships Exploration:** *deductive* techniques start from known effects to seek unknown causes; *inductive* techniques start from known causes to forecast unknown effects. In contrast, *exploratory* techniques relate unknown causes to unknown effects.

Table 2.1 summarizes the characteristics of the dependability analysis techniques according to the criteria mentioned above: the applicability in the lifecycle (second column); the aim (third column), i.e., qualitative (ql.) vs/quantitative (qn.); bottom-up vs/ top-down (fourth column); and, finally, the type of exploration of the cause-effect relationship (fifth column).

In this chapter, we discuss three techniques used for hazard analysis: (1) Fault-Tree Analysis (FTA): across, qualitative and quantitative, top-down, deductive; (2) HAZOP: early, qualitative, bottom-up, exploratory; (3) STAMP/STPA: across, qualitative, top-down/bottom-up, inductive/deductive technique.

2.3.2 Fault-Tree Analysis (FTA)

A Fault Tree comprises logic gates (AND, OR, K-of-M) and external nodes corresponding to component/subsystem faults [4] illustrated in Table 2.2. This technique allows representing and analyzing the component faults whose combined occurrence produces a system failure. It is a top-down technique; accordingly, the representation of a fault tree initiates by selecting a system failure mode (the top event). It ends when either the basic events

Table 2.1 Characteristics of the dependability analysis techniques (adapted from [4]).

Technique	Lifecycle	Aim	Bottom-up/ Top-down	Cause-effect relationship exploration
ETA	across	ql./ qn	bottom-up	inductive
FMEA	across	ql	bottom-up	inductive
FMECA	across	qn	bottom-up	inductive
FTA	across	ql./ qn	top-down	deductive
FFA	early	ql	bottom-up	deductive
HAZOP	early	ql	bottom-up	exploratory
Markov analysis	late	qn	top-down	not applicable
PN	late	ql./ qn	top-down	inductive
PHA	early	ql	bottom-up	exploratory
RBD	across	ql./ qn	top-down	inductive
Truth table	late	ql./ qn	not applicable	inductive
STAMP/ STPA	across	ql	top-down/bottom-up	inductive/ deductive

Table 2.2 Main constructs of fault trees (adapted from [4]).

Symbol	Name	Description
	Event	A fault event that happens as a result of the logical combination of other events.
AND	AND Gate	The output event occurs only if ALL input events occur.
OR	OR Gate	The output event occurs only if (at least) one or more input events occur.
k/n	K-of-M gate	Output event that happens if K or more of the M input events occur.

that trigger such a failure are all identifiedor the desired level of detail is achieved [4]. Therefore, they are suitable for constructing models following a hierarchical approach that contributes to improving the model readability and reusability.

An example of a fault tree, which is a partial view of the "insulin overdose" hazard [8], is illustrated in Figure 2.4. The top event (Insulin overdose) is caused by one of the events at the lower level, which are then connected with an OR gate. To improve the FTA understanding, Martins and Oliveira [8] put tags near the fault tree leaves. These tags identify if the

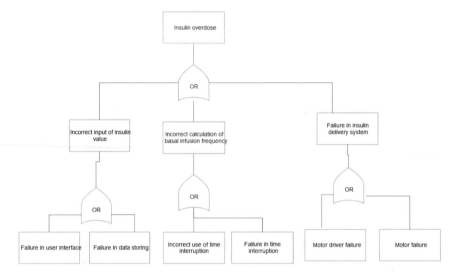

Figure 2.4   Fault tree for the "insulin overdose" hazard with classification tags [8].

failure cause is a software cause (S), an electronic cause (E), or a mechanical cause (M).

2.3.3 HAZOP

HAZard and OPerability study (HAZOP) is a qualitative technique based on guidewords (e.g., none, more of, less of, part of, more than, other) that represent a possible deviation from the normally expected characteristic of a system [4]. Table 2.3 presents some examples of guidewords and their meaning presented in IEC 61882 [11], describing a flow chart of the HAZOP examination procedure.

Driven by the guidewords, the causes of failures and their effects are listed. Each effect is then considered, and actions must be proposed to mitigate it or decrease the probability of the failure cause. The information elicited with HAZOP is represented in a tabular form, such as the worksheet in Figure 2.5.

A multidisciplinary team often carries out HAZOP during the early stages of the system life cycle (i.e., requirement analysis, high-level design) to anticipate hazards [4]. Moreover, it can be used in conjunction with other dependability analysis methods such as fault tree analysis explained in the previous section.

Table 2.3 Examples of deviations and their associated guidewords (adapted from [11]).

Deviation type	Guideword	Example interpretation for process industry
Negative	NO	No part of the intention is satisfied, e.g., no flow.
Quantitative modification	MORE	A quantitative increase, e.g. higher temperature.
	LESS	A quantitative decrease e.g. lower temperature.
Quantitative modification	AS WELL AS	Execution of steps at the same time.
	PART OF	Only some intention is achieved, i.e., only part of an intended fluid transfer occurs.
Substitution	REVERSE	Covers reverse flow in pipes and reverse chemical reactions.
	OTHER THAN	A result other than the original goal is achieved, i.e., transfer of wrong material.
Time	EARLY	Something occurs early relative to clock time, e.g., cooling or filtration.
	LATE	Something takes place late relative to clock time, e.g., cooling or filtration.
Order or sequence	BEFORE	Something occurs too early in a sequence, e.g., mixing or heating.
	AFTER	Something happens too late in a sequence, e.g., mixing or heating.

STUDY TITLE:							SHEET:		
Drawing No.:			REV. No.:				DATE:		
TEAM COMPOSITION:							MEETING DATE:		
PART CONSIDERED:			Transfer line from supply tank A to reactor						
DESIGN INTENT:			Material:	Activity:					
			Source:	Destination:					
No.	Guide word	Element	Deviation	Possible causes	Consequences	Safeguards	Comments	Actions required	Action allocated to

Figure 2.5 Example of a HAZOP worksheet.

2.3.4 STAMP/STPA

STAMP is a causality model proposed by Nancy Leveson [1] that relies on three main concepts: *safety constraints, safety hierarchical control structures*, and *process models*. Hence, instead of defining safety management in preventing component failures, STAMP proposes creating a safety control structure that will enforce the behavioral safety constraints and ensure their continued effectiveness as changes and adaptations occur over time [1]. STPA [1] is an approach to hazard analysis that can be used at any stage of the system life cycle, and it defines two main steps [1]:

Step 1: Identify the potential for inadequate control of the system that could lead to a hazardous state.

Step 2: Determine how each potentially hazardous control action identified in step 1 could occur.

The information provided in the first step of STPA can be used to detect the necessary constraints on component behavior to avoid the identified system hazards, which are specified as safety requirements. In the second step, the information required by the component to properly implement the constraint is identified, and additional safety constraints and information necessary to eliminate or control the hazards in the design or properly design the system in the first place [1]. According to Leveson [1], a table or other types of representation may be used to record the HA results.

2.4 Step 2: Hazards Evaluation

In the *hazards evaluation* step, it is estimated the **probability** of the identified hazards and their **severity**. Usually, it is not possible to perform this accurately due to the absence of enough quantitative data available for statistical analysis. In this context, common qualitative values are used for the **severity**:

- *Not Significant*: Minor injuries or discomfort. No medical treatment or measurable physical effects;
- *Minor*: Injuries or illnesses requiring medical treatment. Temporary impairment;
- *Moderate*: Injuries or illness requiring hospital admission;
- *Major*: Injury or illness resulting in permanent impairment;
- *Severe*: Fatality.

  Common qualitative values used for **probability** are:

- *Almost Certain*: Expected to occur regularly under normal circumstances;
- *Likely*: Expected to happen at some time;
- *Possible*: May occur at some time;
- *Unlikely*: Not likely to occur in regular circumstances;
- *Rare*: It could happen, but probably never will.

Table 2.4 An example of a template to document hazard analysis.

Hazard analysis					Page	of
Company					Date:	_/_/_
System						
Number	Hazard	Cause	Probability	Severity	Risk	Aceitability

An example of a template that could be used to document the hazard analysis is presented in Table 2.4.

The values of *risk* and *aceitability* are defined after performing the risk analysis described in the next step.

2.4.1 Step 3: Risk Analysis

The risk analysis step determines the risk of each identified hazard considering the combination of the probability and the severity. This analysis is performed considering a risk analysis matrix, such as the one illustrated in Figure 2.6. After the determination of the risks, we should determine their Aceitability. According to the literature, we can categorize them as:

- **Intolerable**: These are those that threaten human life and should never arise or result in an accident. In the case of the insulin pump, an unacceptable risk is an insulin overdose.
- **Tolerated** (as low as reasonably practical – ALARP): Those whose consequences are less severe or severe but have a very low probability of occurrence. They should be mitigated unless risk reduction is impracticable or excessively costly. An ALARP risk for an insulin pump may be hardware monitoring system failure. The consequences are, at worst, short-term insulin underdose. This is a situation that would not cause a severe accident.
- **Acceptable**: These are those in which the associated accidents usually result in minor damage. The risk consequences are acceptable and should be made at no additional cost to reduce the probability of risks. An acceptable risk in an insulin pump may be the risk of an allergic reaction arising in the user; This usually only causes skin irritation. It would not be worth using more expensive special materials on the device to reduce this risk.

In the next section, we detail the last step performed in safety analysis: the specification of safety requirements.

			Severity				
			Minor injuries or discomfort. No medical treatment or measurable physical effects	Injuries or illnesses requiring medical treatment. Temporary impairment.	Injuries or illness requiring hospital admission.	Injury or illness resulting in permanent impairment.	Fatality
			Not Significant	Minor	Moderate	Major	Severe
Probability	Expected to occur regularly under normal circumstances.	Almost Certain	Medium	High	Very High	Very High	Very High
	Expected to occur at some time.	Likely	Medium	High	High	Very High	Very High
	May occur at some time.	Possible	Low	Medium	High	High	Very High
	Not likely to occur in normal circumstances.	Unlikely	Low	Low	Medium	Medium	High
	Could happen, but probably never will.	Rare	Low	Low	Low	Low	Medium

Figure 2.6   Risk analysis matrix.

2.5 Safety-related Requirements Specification

The purpose of the safety-related requirements specification is to identify the Safety-Related Requirements that determine how risks should be managed and ensure no accidents/incidents occur. According to Firesmith [12], the requirements can be classified as:

- **Safety Requirements**: is a combination of a safety criterion and a minimum threshold on a safety measure. They directly specify system safety, and they are considered a kind of quality requirement [12]. Examples: at the insulin pump, the difference between the scheduled infusion and the delivered infusion should not exceed 0.5%; The system should not cause more than 5 accidental damages per year;

- **Safety-Significant Requirements**: they are non-safety primary mission requirements with significant safety ramifications. They can be classified as a subset of non-safety requirements: Functional Requirements, Data Requirements, Interface Requirements, Non-safety Quality Requirements, and Constraints [12]. Examples of such requirements are "Requirements for controlling elevator doors such as Keep doors closed when moving"; and "Not crush passengers."
- **Safety System Requirements or Safety Functional Requirements:** These are requirements for safety systems or subsystems. An example of such requirements is "The Fire Detection, and Suppression System shall detect smoke above X ppm in the weapons bay within 5 seconds" [12].
- **Safety Constraints:** they are any constraint primarily intended to ensure a minimum level of safety. For example: "Oils and hydraulic fluids shall be flame retardant, except as required for normal lubrication" [12].

The determination of such requirements relies on the techniques to obtain safety described in the next section.

2.5.1 The Means to Obtain Safety

After identifying the hazards and the associated risks, we must specify the safety requirements aiming to eliminate or prevent hazards [1]. The technique being used depends on the hazard *aceitability* previously documented and the number of resources available. The safety requirements can be derived from the following set of four risk reduction strategies [3]:

1. *Fault avoidance:* it consists of using techniques to prevent, *by construction*, the occurrence of faults. Therefore, the system is designed so that hazards cannot occur.
2. *Fault detection and removal* aim to minimize, using *verification and validation* techniques, the presence of faults. Hence, the system is designed to detect and neutralize hazards before they result in an accident. In this strategy, static and dynamic analysis can be performed as well as fault injection.
3. *Fault-tolerance:* it has the goal of providing, for *redundancy*, specification-compliant service even in the presence of hazards. Fault tolerance does not eliminate the need for using avoidance or removal techniques.

4. *Fault forecasting:* it involves minimizing, *by qualitative or quantitative evaluations or using estimates*, the presence, future incidence, and the likely consequences of faults.

Usually, a combination of risk reduction strategies is used. For example, in a chemical plant control system, the system will include sensors to detect and correct excess pressure in the reactor. However, it will also include an independent safety system, which opens a relief valve if the risk of dangerously high pressure is detected. In the following sections, we describe some safety requirements specification techniques.

2.5.2 Model-driven Approaches

Model-driven development (MDD) is defined as the use of models for developing software [4]. A model-driven approach to conducting hazard analysis on use cases [6] requirements representations is described by Allenby and Kelly [5]. Using the approach, it is possible to specify safety requirements from use cases. Their approach consists of three main steps: 1. Representation of core functionality by the identification of core use cases; 2. Identification and documentation of the scenarios for each core use case; 3. Decomposition and allocation of functionality across communicating subsystems. In the end, functional requirements at each development level under consideration will be represented in use case diagrams.

Thramboulidis and Scholz [16] present an approach to integrate safety analysis with the 3+1 SysML view model, a SysML-based architectural approach for mechatronic system development. The first activity is to apply Hazard Analysis based on the requirements composed of SysML requirements diagrams and essential use cases. In the second phase of safety analysis, a solution-dependent hazard analysis that results, among others, in defining required Safety Integrity Levels (SILs) for the system components and their assigned functions is performed.

2.5.3 Textual-driven Approaches

Safety requirements documentation can also use the textual format. Fu, Bao e Zhao [17] define embedded software generic safety requirements description templates to perform safety analysis based on controlled natural language and requirements description templates. An example of a failure detection description template is presented in Figure 2.7.

Figure 2.7   Failure detection description template of generic safety requirements [17].

The templates include safety requirements structural elements description templates and safety requirements sentence pattern description templates based on obtained structural elements, failure modes, safety strategies.

2.5.4 Model-driven Approaches Combined with Natural Language Specification

The combination of model-based safety analysis and text expressed by natural language is also adopted by the literature to improve the understandability and result in more detailed safety requirements. Martins and Oliveira [8] proposed a protocol to help requirements engineers to derive safety functional requirements from FTA. The requirements are expressed in the style "Should" and "Should Not" Requirements. Columns (3) and (4) in Table 2.5 present some "should" and "should not" requirements obtained from the fault tree of Figure 2.4.

2.5.5 Ontological Approach to Elicit Safety Requirements

Ontologies have been used for requirement elicitation [13] as well as for safety requirements elicitation [14][15]. Provenzano et al. define a heuristic Safety Requirements Elicitation (SARE) approach to elicit safety requirements considering knowledge obtained in the previous safety analysis. This knowledge comprehends the causes, sources, and consequences of hazards. Their approach consists of three activities: (1) overcome an object's weakness; (2) change, add or remove an object's role; (3) cut off existing relation; and it relies on a hazard ontology to guide to derive the safety requirements more appropriate to mitigate the associated hazards.

The work of Zhou et al. [15] uses conceptual ontologies to 1) systematically organize the knowledge of the system operating environment and 2) simplify the elicitation of environmental safety requirements. The ontologies organize the environment knowledge in terms of relevant environment concepts, relations among them, and axioms in their approach. Environmental assumptions are captured by instantiating the environment ontology. The approach consists of four steps: (1) Preparation, (2) Ontology

Table 2.5   A partial view of safety functional requirements in should/should not style [8].

(1) Safety Requirements	(2) Failure causes	Safety Functional Requirements	
		(3) "Should" Requirements	(4) "Should Not" Requirements
Incorrect input of insulin value	Failure in user interface (S)	1. The system **should** delimit the insulin value range during the specification of basal infusion profile. 2. The input of insulin value to basal infusion profile specification **should** be done using up and down buttons.	1. The system **should not** allow the user to choose insulin value out of the safety specified range for basal infusion profile.
	Failure in data storing (S)	3. The system **should** store at least 3 basal infusion profiles, each one specifying 24 insulin values (one by hour).	

Definition, (3) Rule Definition, and (4) Safety Requirements Elicitation. Finally, an ontological reasoning mechanism is also provided to support the elicitation of safety requirements from the captured assumptions.

2.6 Conclusions

Safety is a system property representing its ability to avoid failures that could be catastrophic to users or the system's environment. In this chapter, we discussed the foundations of safety engineering by explained the F-E-F (faults, errors, and failures) chain and safety-related concepts such as Accident, Safety Incident, Hazard, Environmental or operational conditions (Context), Harm, Hazard Severity, and Probability, Risk, and Safety Functional Requirement.

We also presented the safety and hazard analysis method, which has four steps: 1- Hazards Identification, 2- Hazards Evaluation, 3- Risk Analysis, and 4- Safety-Related Requirements Specification by presenting examples of methods and techniques that could be used in these steps.

From the safety analysis results, safety requirements are derived, and they should be incorporated in the system design before implementation, aiming to develop safer systems by reducing the occurrence of hazards from the beginning of the system development process. Such an approach of integrating requirements and safety engineering highlights the relationship between these two areas.

Acknowledgments

We would like to thank Universidade Federal of Pernambuco (UFPE), Brazil.

References

[1] N. Leveson. Engineering a safer world: Systems thinking applied to safety. Mit Press, 2011.

[2] J. Vilela, J. Castro, L. E. G. Martins, T. Gorschek. Safety Practices in Requirements Engineering: The Uni-REPM Safety Module. IEEE Transactions on Software Engineering, 2018.

[3] A. Avizienis, J. C. Laprie, B. Randell. Fundamental concepts of dependability. University of Newcastle upon Tyne, Computing Science, 2001, pp. 7-12.

[4] S. Bernardi, J. Merseguer, and D. Petriu. Model-driven dependability assessment of software systems. Heidelberg: Springer, 2013.

[5] K. Allenby, T. Kelly. Deriving safety requirements using scenarios. In: Proceedings Fifth IEEE International Symposium on Requirements Engineering, 2001, pp. 228-235.

[6] G. Booch. The unified modeling language user guide. Pearson Education India, 2005.

[7] J. Vilela, J. Castro, L. E. G. Martins, T. Gorschek, C. Silva. Specifying Safety Requirements with GORE languages. In: Proceedings of the 31st Brazilian Symposium on Software Engineering, 2017, pp. 154-163.

[8] L. E. G Martins, T. A. Oliveira. A case study using a protocol to derive safety functional requirements from fault tree analysis. In: International Requirements Engineering Conference (RE), 2014, pp. 412-419.

[9] I. Sommerville. Software engineering 9th Edition, 2011.

[10] J. Vilela, J. Castro, L. E. G. Martins, T. Gorschek Safe-RE: a safety requirements metamodel based on industry safety standards. In: Proceedings of the XXXII Brazilian Symposium on Software Engineering, 2018, pp. 196-201.

[11] IEC-International Electrotechnical Commission. IEC 61882, 2001.

[12] D. Firesmith. Engineering safety-related requirements for software-intensive systems. In Proceedings of 27th International Conference on Software Engineering, 2005, pp. 720-721.

[13] D. Dermeval, J. Vilela, I. I. Bittencourt, J. Castro, S. Isotani, P. Brito, and A. Silva; Applications of ontologies in requirements engineering: a systematic review of the literature. In:Requirements Engineering Journal, 21(4), 2016, pp.405-437.

[14] L. Provenzano, K. Hänninen, J. Zhou, and K. Lundqvist. An Ontological Approach to Elicit Safety Requirements. In: 24th Asia-Pacific Software Engineering Conference (APSEC), 2017, pp. 713-718.

[15] J. Zhou, K. Hänninen, K. Lundqvist, Y. Lu, L. Provenzano, and K. Forsberg. An environment-driven ontological approach to requirements elicitation for safety-critical systems. In 23rd International Requirements Engineering Conference (RE), 2015, pp. 247-251.

[16] K. Thramboulidis and S. Scholz. Integrating the 3+ 1 SysML view model with safety engineering. In 2010 IEEE 15th Conference on Emerging Technologies & Factory Automation (ETFA 2010), pp. 1-8.

[17] R. Fu, X. Bao, and T. Zhao. Generic safety requirements description templates for the embedded software. In 2017 IEEE 9th International Conference on Communication Software and Networks (ICCSN), 2017, pp. 1477-1481.

3

Integrating New and Traditional Approaches of Safety Analysis

L. E. G. Martins

Department of Science and Technology, Federal University of São Paulo, São José dos Campos, Brazil
E-mail: legmartins@unifesp.br

Abstract

Traditional approaches as FMEA and FTA still are very used during the process of safety analysis. Practitioners are used with them, and they are resistant to move to new approaches. However, traditional approaches are no longer sufficient to deal with the increasing complexity of safety-critical systems. New approaches are needed; STAMP/STPA is emerging as a promising approach. In this chapter, we discuss integration possibilities between new and traditional safety analysis approaches. We briefly present FMEA and FTA as representative of traditional approaches; and STAMP/STPA as representative of a new safety analysis approach. We do an exercise of mapping FMEA and STPA steps and try to identify possibilities to use both approaches in a complementary way.

Keywords: Safety Analysis, Safety Requirements, FMEA, STAMP.

3.1 Introduction

Safety-Critical Systems (SCS) are increasingly present in the daily lives of modern societies, which are becoming heavily dependent on these systems. SCS are technology-based man-made systems where any defects or failures may lead to accidents that endanger human life or damage the environment or property [9–11]. SCS are present in avionic systems, automotive systems, industrial plant control (chemical, oil & gas, nuclear), medical devices, railroad control, defense, and aerospace systems, among others [10, 12].

In this chapter, we discuss integration possibilities between new and traditional safety analysis approaches. We briefly present FMEA and FTA as representative of traditional approaches; and STAMP/STPA as representative of a new safety analysis approach. This chapter is organized as follows: Section 3.2 presents background and related work, in section 3.3 we briefly describe FMEA and FTA, in section 3.4 we briefly describe STAMP/STPA, in section 3.5 we present some ideas of how FMEA and STPA may be used in an integrated way, and in section 3.6 we conclude the chapter.

3.2 Background and Related Work

In this section, we present a background related to safety analysis. To help the reader, we provide definitions for some terms and expressions used along with this chapter. A brief discussion of related work to the integration of new and traditional approaches of safety analysis is provided in section 2.2.

3.2.1 Background

In order to help the reader better understand the discussion along with this chapter and make clear the adopted terms, we provide definitions for some terms and expressions related to safety analysis in the context of requirements engineering and safety-critical systems. We present the following definitions [8], organized in alphabetical order:

- **Accident**. An undesirable (negative) event involving damage, loss, suffering, or death [5, 6].
- **Approach**. In the context of this chapter, we are interested in the following types of approaches: technique, model, framework, method, process, methodology, or tool to elicit, model, specify or validate hazards and safety requirements for safety-critical systems.
- **Functional Safety Requirement**. The requirement to prevent or mitigate the effects of failures identified in safety analysis [4].
- **Hazard**. A system state that might, under certain environmental conditions, lead to a mishap. Hence, a hazard is a potentially dangerous situation that may lead to an accident [2, 5].
- **Safety**. Firesmith defines safety as "the degree to which accidental harm is properly addressed (e.g., prevented, identified, reacted to, and

adapted to)" [3]. According to Leveson, "safety must be defined in terms of hazards or states of the system that when combined with certain environmental conditions could lead to a mishap." [5, 7]

- **Safety Analysis**. A systematic analysis of the architecture, design, installation, and operation of the system to ensure that safety requirements are met. During this analysis process, all hazards and their effects on the system should be addressed [18].
- **Safety Requirement**. A requirement that describes the constraints or actions to support and improve system safety. Firesmith defines the safety requirement as "any quality requirement that specifies a minimum, mandatory amount of safety in terms of a system-specific quality criterion and a minimum level of an associated metric." [3]
- **Safety-Critical**. According to Medikonda and Panchumarthy, "those software or system operations that, if not performed, performed out of sequence, or performed incorrectly could result in improper control functions, or lack of control functions required for proper system operation, that could directly or indirectly cause or allow a hazardous condition to exist" [3].

3.2.2 Related Work

Sulaman et al. [20] performed a case study making a qualitative comparison between FMEA and STPA. Both methods were applied to the same forward collision avoidance system to compare the effectiveness of the methods and to investigate what are the main differences between them. The main results from this case study are the following [20]:

- Almost all types of hazards that were identified in the study were found by both methods;
- Both methods found hazards classified as component interaction, software, component failure, and system type;
- With regard to component failure hazards, FMEA identified more component failure hazards than STPA;
- With regard to software error type hazards, STPA found more hazards than FMEA unique hazards;
- With regard to component interaction error type hazards, STPA found some hazards; however, FMEA did not find any unique hazards;
- With regard to system type error hazards, FMEA found slightly more hazards than STPA;

- Both FMEA and STPA consider system decomposition (FMEA decomposes and STPA considers the whole system for analysis), identification of potential failures, their causes and effects, and definition of countermeasures. But STPA does not consider risk assessment in terms of risk priority number calculation and assignment of the application function to each subsystem.

The final conclusions reported by Sulaman et al. [20] are "the methods have different focuses. FMEA especially takes the architecture and complexity of components into account, whereas STPA is stronger in finding causal factors of identified hazards. It can be concluded that, in this study, there was no type of hazard that was not found by any of the methods, which means that it is not possible to point out any significant difference in the types of hazards found. However, it can be observed that none of the methods in this study was effective enough to find all identified hazards, which means that they complemented each other well in this study."

3.3 Traditional Approaches

In this section, we briefly present two traditional approaches used during safety analysis: failure mode and effect analysis; and fault tree analysis.

3.3.1 FMEA: Failure Mode and Effect Analysis

In the last years, the safety-critical system developers have noted the need for integration of safety engineers and requirement engineers to increase the interaction among the participants of these teams. Such interaction has the potential to bring benefits and gains in the quality requirements as well as to improve the safety level of the SCS [9, 11]. The approximation of the safety and requirement engineers has boosted the interchange of techniques that traditionally were used in an isolated way. One technique that deserves attention in the interaction process between safety analysis and requirements specification is the FMEA (Failure Mode and Effect Analysis).

FMEA was formalized in 1949 by the United States Army. This technique aims to identify and prioritize failures according to their impacts on the mission success and safety of the people and equipment [1, 14]. Currently, FMEA is extensively used in several domains in the SCS industry, such as avionics, automotive, defense, medical devices, industrial plants, food, among others [15]. FMEA has passed for improvements over the years, offering effective help for [16]: (a) to identify and understand the failure modes and their causes, as well as the failure effects over the system or their users, considering a specific product or process; (b) to evaluate the associated risks to their causes, effect modes and identified causes; (c) to identify, prioritize and execute corrective actions. The main motivation of the use of FMEA is to preview and avoid accidents in such a way that products and processes be safe, putting the associate risks at a low level.

There are several types of FMEA, among which the main are [1]: system FMEA, design FMEA, process FMEA, and service FMEA. A system FMEA, also called concept FMEA, is composed of a series of steps to include conceptual design, detailed design, development, test, and evaluation. The results of the system FMEA will be used as input for the design FMEA. An effective system FMEA should be carried out along with a system engineering process or a product development process. A design FMEA, also called product FMEA, is a systematic method of identifying potential or known failure modes and offering corrective actions before the first production run occurs [1]. Design FMEA is managed based on a series of steps, which include component and subsystems development and integration. The outcome of the design FMEA will be used as input for the process and service FMEA. The process FMEA normally is managed to take into account elements as labor, machine, method, material, measurement, and environmental considerations. The goal of a service FMEA is to define proper solutions in response to quality and productivity as defined by the design specification and customer requirements.

FMEA is among the safety analysis approaches considered traditional. According to a systematic literature review published in 2016 [15], FMEA was indicated as the most used approach for safety analysis. Other traditional approaches that are largely used in the SCS industry are FTA (Fault Tree Analysis) and HAZOP (Hazard and Operability Study) [17]. According to Leveson [10], these traditional approaches, which are based on the event chain analysis, accomplished an important role to support the risk and safety analysis over the last decades. However, they are no longer adequate to handle the increased complexity of the current SCS. The main problems that are not addressed by the traditional approaches are the following [10]: (a) the introduction of new technology has the potential to create new types of hazard that can lead to accidents (the replacement of electromechanical to computerized components is a classic case); (b) the increased interrelationship and coupling among different components and subsystems; (c) the inadequate communication among people, systems and

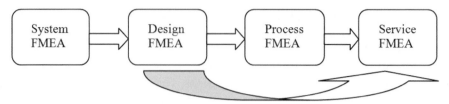

Figure 3.1   Main types of FMEA and their relationships.

machines are becoming an increasingly important factor in accidents; (d) an accident usually is a complex process involving a whole socio-technical system, the traditional approaches are not able to describe such a process in a proper way.

3.3.2 FTA: Fault Tree Analysis

FTA is considered a deductive analytical technique used in the context of safety analysis for complex dynamic systems. FTA was developed by Bell Telephone Company in 1961. Since then, FTA is largely used in many industry domains and companies [1]. FTA uses a fault tree to model the cause and effect relationships that may occur between an undesired event (a hazard) and several contributing causes. Using the standard symbols to build the fault tree (see Figure 3.2) is possible to represent various combinations of the faulty and normal events, which may lead to a top undesired event (a system hazard).

The fault tree shows the logical paths from the top undesired event, which can be a system hazard in analysis, to the possible causes represented at the bottom of the tree (leaves). Usually, FTA is a complementary model to the FMEA. In [1], Stamatis points out several benefits of using FTA, among them:

- To help in visualizing the safety analysis;
- To help in identifying the reliability of higher-order (level) assemblies or the system;
- To determine the probability of occurrence for each of the root causes;
- To provide documented evidence of compliance with safety requirements;
- To assess the impact of design changes and alternatives;
- To provide options for qualitative, as well as quantitative, system reliability analysis;
- To allow the safety engineer to concentrate on one particular system failure at a time;

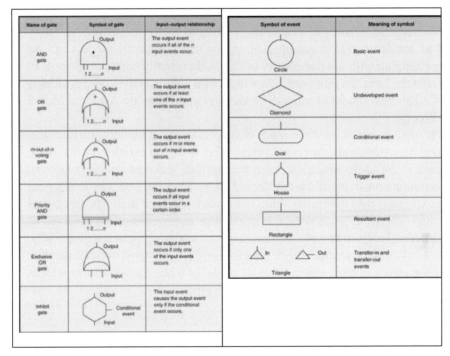

Figure 3.2 Main standard symbols used to build a fault tree (extracted from [1]).

- To provide the safety and system engineers with insight into system behavior;
- To isolate critical safety failures; and
- To identify ways that failure of the product can lead to an accident.

3.4 New Approaches

In this section, we introduce two new approaches that can contribute to improving the safety analysis of safety-critical systems: STAMP and STPA. Both approaches were developed by Leveson [10] and can be used in an integrated way.

3.4.1 STAMP

In the last decade, traditional models for safety and accident analysis in SCS have been questioned for effectiveness [10]. Such (traditional) models are based on the analysis of failure event chains, where each failure can result in a subsequent failure within a string, leading to a "domino effect" and may lead to accidents [13]. Leveson [10] argues that traditional models can no longer keep up with the complexity of the SCS currently being developed. In this context, Leveson proposed a new model for safety analysis and accidents in SC. This new model is called STAMP: System-Theoretic Accident Model and Process.

STAMP is based on three main concepts: safety constraint, hierarchical control structure, and process models. The safety constraint is the basic concept in STAMP, which can take the form of a design, implementation, or operation constraint of the developing SCS. The safety constraint is a goal to be achieved by a controller (human or machine). The hierarchical control structure maps the relationships between the actors/controllers involved at the different levels of control, design, development, and operation of the SCS. Process models are abstract representations of SCS that need to be controlled. These models can represent the mental model of a human operator as well as the control logic of an automated controller [10].

3.4.2 STPA

The System-Theoretic Process Analysis (STPA) is a new approach to hazard analysis based on the STAMP causality model proposed by Leveson [10]. According to Leveson, *"the primary reason for developing STPA was to include the new causal factors identified in STAMP that are no handled by the older techniques"* [10]. Examples of older or traditional techniques include FTA, ETA, FMEA, HAZOP, among others. The intention of STPA is to become a broader hazard analysis technique, covering aspects such as design errors, software flaws, component interaction accidents, human decision-making errors, and social, organizational, and management factors contributing to accidents.

STPA is divided into two steps [10]:

(I) **Identification of the potential for inadequate control of the system.** Inadequate control may lead to hazardous states of the system, which can occur because:
 a. A control action required of safety is not provided or not followed;
 b. An unsafe control action is provided;
 c. A potentially safe control action is provided too early or too late; and
 d. A control action required for safety is stopped too soon or applied too long.

(II) **Determination of how each potentially hazardous control action identified in step (I) could occur.** This step is composed of two main processes:

  a. For every hazardous control action identified, evaluate the parts of the control loop. Design controls and mitigation actions if they do not exist, or evaluate existing actions if the analysis is being carried out on an existing design. If there are multiple controllers, identify potential conflicts and coordination issues.

  b. Analyze how the designed controls could degrade over time and build in protection.

As an example of how to use STPA, we adopt an insulin infusion pump system. In Figure 3.3 , we show the control structure for a hazard "Stepper motor stopped working" in the context of an insulin infusion pump. In this figure, the components of the system are shown along with the control information that each component can send and receive. The insulin pump controller sends the number of steps to be carried out by the stepper motor. The sensor measures the traveled distance by the plunger each minute, which is sent to the insulin pump controller. The controller verifies the traveled distance along the time. If the distance is the same for more than one hour, then the alarm sound and alarm message are turned on.

Hazard: Stepper motor stopped working

System Safety Constraint: The alarm must be on whenever the stepper motor is not working properly.

Functional Requirements: (1) Detect when the stepper motor stopped working; (2) Turn on the alarm sound; (3) Show an alarm message on display.

We performed the first step of STPA in the insulin infusion pump system to determine the potential of inadequate control leading to a hazard. In table 3.1, we show the results of this part of the hazard analysis.

The next step of STPA (step II) is the determination of how each potentially hazardous control action identified in step I could occur. As showed in Table 3.1, we identified three potentially hazardous situations: (1) Sensor does not detect that the plunger is stopped; (2) Sensor send traveled distance = 0 when plunger moved; and (3) Sensor provides traveled distance too late. Situation (1) could occur because of a malfunction of the sensor; situation (2) could occur because of a malfunction of the sensor or when something is in between the sensor and plunger; Situation (3) could occur

Table 3.1 Identification of the potential for inadequate control (step I of STPA).

Control Action	Not provided or not followed	The unsafe control action is provided	Provided too early or too late	Stopped too soon or applied too long
Detection that the stepper motor has stopped	The sensor does not detect that the plunger is stopped	Sensor send traveled distance = 0 when plunger moved	The sensor provides traveled distance too late	N/A

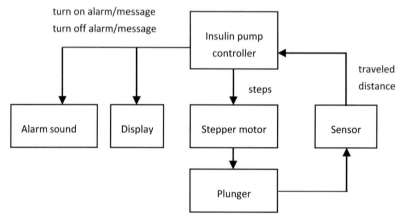

Figure 3.3 Control structure for insulin infusion pump (partial representation).

when the synchronization between the sensor and infusion pump controller is misadjusted.

3.5 Integration Between New and Traditional Approaches

Traditional approaches as FMEA and FTA still are very used during the process of safety analysis (as discussed in section 3.1 [15]). Practitioners are used with them, and they are resistant to move to new approaches. However, traditional approaches are no longer sufficient to deal with the increasing complexity of SCS [10]. New approaches are needed, and STAMP/STPA is emerging as a promising approach. Nevertheless, the transfer of new technology to the industry setting cannot be performed abruptly [19]. For SCS professionals to adopt a new way of hazard/safety analysis of their systems, they will need to identify real benefits in the proposed new approaches.

Table 3.2 Mapping of the analysis steps for FMEA and STPA [20].

FMEA	STPA	Mapping Comments
Step 1: Decomposition of the system to be analyzed into subsystems and components	*Step 1*: Acquisition of functional control diagram of the system to be analyzed as a whole, and identification of some high-level system hazards to start with	*Map-A*: Step 1 of both methods are mapped as the same step in the analysis process because FMEA is based on reliability theory (decomposition required), and STPA is based on system theory (system required as a whole)
Step 2: Assigning the application function to each subcomponent and subsystem	N/A	*Map-B*: This step of FMEA does not map to any STPA step
Step 3: Determine and analyze: —potential failure modes —causes of failure —failure effects that can lead the system to a hazardous state	*Step 2*: Identify the potential inadequate control commands or events (potential hazards) *Step 3*: Determine how each potential hazardous control action (potential hazards) identified in step 2 could occur (causal factors of identified potential hazards)	*Map-C*: Step 3 of FMEA is mapped to step 2 of STPA, which consists of the identification of potential failures (or hazards), their causes and effects
Step 4: Evaluate risk and calculate risk priority number (RPN)	N/A	*Map-D*: This step of FMEA does not map to any STPA step
Step 5: Specify defect avoidance or risk mitigation measures	*Step 4*: Design controls and countermeasures if they do not already exist or evaluate existing	*Map-E*: Step 5 of FMEA and step 2 of STPA are mapped to each other because they are both about designing and evaluating countermeasures

Switching from the traditional to the new often faces resistance and is no trivial task. This exchange involves risks and cultural changes [19] and usually entails costs for companies at first.

Sulaman et al. [20] performed a case study comparing FMEA and STPA (aforementioned in section 2.2). The results from this case study point outs interesting ideas about a possible integration between these two methods in order to complement each other. Table 3.2 was extracted from the work of

Sulaman et al. [20] and presents a mapping of the analysis steps for FMEA and STPA.

According to the results from the case study performed by Sulaman et al. [20], FMEA and STPA have different focuses. FMEA is based on the reliability theory, whereas STPA is based on the system theory. It seems that FMEA is more appropriated than STPA for identifying component failure hazards. According to Sulaman et al. [20], "this shows the basic philosophy behind both methods: FMEA focuses more on components, their failures, and risk mitigation measures, whereas STPA focuses on the delivery of control commands and their feedbacks."

Another difference mentioned by Sulaman et al. [20] "where STPA seems to outperform FMEA is finding causal factors of identified hazards."

3.6 Conclusion

In this chapter, we started a discussion about the integration between new and traditional safety analysis approaches. FMEA is a traditional approach still much used by SCS practitioners during hazard and safety analysis [8]. STPA is a new hazard analysis approach proposed by Leveson [10], which is calling the attention of SCS practitioners a promising method. STPA supporters believe that this new method could replace traditional safety analysis approaches such as FMEA. However, the adoption of new technology by industry practitioners does not usually occur quickly.

In addition, STPA and FMEA have different characteristics that can bring benefits to practitioners if used in a complementary manner. Therefore, rather than considering replacing one approach with another, researchers in the field should look for ways to use both approaches in an integrated manner. Such integration possibility can be a promising research topic.

Acknowledgments

This work was partially funded by the Federal University of São Paulo (UNIFESP) and the Brazilian research agency FAPESP – Fundação de Amparo à Pesquisa do Estado de São Paulo - under grant agreement no. 2019/09396-0.

References

[1] Stamatis, D H. (2003). Failure Mode and Effect Analysis: FMEA from Theory to Execution. 2nd Edition, American Society for Quality, Quality Press, Milwaukee.

[2] Medikonda, B. S. and Panchumarthy, S. R. (2009). A Framework for Software Safety in Safety-Critical Systems. In *SIGSOFT Software Engineering Notes*, vol. 34, no. 2. March (pp. 1-9).

[3] Firesmith, D. (2004). Engineering Safety Requirements, Safety Constraints, and Safety-Critical Requirements. *Journal of Object Technology*, vol. 3, no. 3, march/april (pp. 27-42).

[4] Allenby, K. (2001). Deriving Safety Requirements Using Scenarios. In *Proceedings Fifth IEEE International Symposium on Requirements Engineering*, (pp. 228–235).

[5] Johnson, C. (2002). Forensic software engineering: are software failures symptomatic of systemic problems? *Safety Science*, *40*(9), (pp. 835–847). doi:10.1016/S0925-7535(01)00086-8.

[6] Berry, D. M. (1998). The Safety Requirements Engineering Dilemma. In Proceedings of the Ninth International Workshop on Software Specification and Design (pp. 147-149).

[7] Leveson, N. G. (1986). Safety: Why, What, and How. ACM Computing Surveys (CSUR), 18(2), (pp. 125–163).

[8] Martins, L. E. G. and Gorschek, T. (2016) "Requirements Engineering for Safety-Critical Systems: A Systematic Literature Review", Information and Software Technology, Vol. 75, (pp.71–89).

[9] Hatcliff, J., Wassyng, A., Kelly, T., Comar, C., and Jones, P. (2014). Certifiably safe software-dependent systems: challenges and directions. In *Proceedings of the on Future of Software Engineering - FOSE*, (pp. 182–200).

[10] Leveson, N. G. (2011). Engineering a Safer World: Systems Thinking Applied to Safety. The MIT Press.

[11] Heimdahl, M. P. E. (2007). Safety and Software Intensive Systems: Challenges Old and New. In *FoSE 2007: Future of Software Engineering* (pp. 137–152).

[12] Nair, S., de la Vara, J. L., Sabetzadeh, M., and Falessi, D. (2015). Evidence management for compliance of critical systems with safety standards: A survey on the state of practice. *Information and Software Technology*, *60*, (pp. 1–15).

[13] Perneger, T. V. (2005). The Swiss cheese model of safety incidents: Are there holes in the metaphor? *BMC Health Services Research*, 5, 1–7. http://doi.org/10.1186/1472-6963-5-71

[14] Carlson, C. S. (2014). Understanding and Applying the Fundamentals of FMEAs. Tutorial Notes AR&MS (pp. 1-35).

[15] Martins, L. E. G. and Gorschek, T. (2016). Requirements Engineering for Safety-Critical Systems: A Systematic Literature Review, Information and Software Technology, Vol. 75, July 2016, (pp.71–89).

[16] Carlson, C. S. (2012). Effective FMEAs: Achieving Safe, Reliable, and Economical Products and Processes Using Failure Mode and Effect Analysis. Wiley.

[17] Martins, L. E. G. and Gorschek, T. (2017). Requirements Engineering for Safety-Critical Systems: Overview and Challenges. IEEE Software, v. 34, (pp. 49-57).

[18] Wang, P. (2017). System Safety Assessment. Chapter 8 In Civil Aircraft Electric Power System Safety Assessment. Butterworth-Heinemann, (pp. 217-238).

[19] Ivarsson, M. and Gorschek, T. (2009). Technology Transfer Decision Support in Requirements Engineering Research: A Systematic Review of REj. Requirements Engineering Journal, vol. 14, no. 3, (pp. 155-175).

[20] Sulaman, S. M.; Beer, A.; Felderer, M; and Höst, M. (2019). Comparison of the FMEA and STPA safety analysis methods – a case study. Software Quality Journal, vol. 27, no. 1, (pp. 349-387).

4

Agile Requirements Engineering

Jéssyka Vilela

Universidade Federal de Pernambuco, Brazil
E-mail: jffv@cin.ufpe.br

Abstract

Software development methodologies are followed to develop safety-critical systems. This chapter discusses the key features of agile software development methodologies in the definition of safety requirements. We also explain the main phases of requirements engineering and agile approaches proposed to each of these phases. We provide a discussion about a balance between conservative and agile methodologies during Safety-Critical Systems (SCS) development. Finally, a case study from the pharmaceutical and another from the avionics domain are presented.

Keywords: Requirements Engineering, Agile, Traditional software development, Software processes, Scrum, XP.

4.1 Introduction

Companies follow software development processes or software lifecycles to develop a system. They consist of a set of stages that defines activities to be performed to construct a system. These processes can be classified in development paradigms such as traditional, agile, and emergent software processes.

In the traditional paradigm, the stages required to construct the software are executed sequentially, and each phase begins only when the previous one is finished. We can mention the waterfall model and the V-model [2].

The main characteristic of the agile paradigm is that the stages are performed iteratively in which a small set of features are implemented in each iteration. XP (eXtreme Programming) [6] and Scrum [5] are examples of software processes that belong to this paradigm. Finally, the development based on reuse and model-driven development (MDD) are examples of emergent software development processes in which a set of principles are followed.

Regardless of the software development paradigm adopted, they have a common set of four activities considered fundamental to software engineering [2]: software specification, software design and implementation, software validation, and software evolution. In this chapter, our emphasis will be on software specification, one activity of Requirements Engineering (RE), in agile software processes.

Although there are challenges in adopting agile methods to develop an SCS [25], it is discussed in the literature that agile software processes can be followed to build a safety-critical system [19][24][25]. A survey with 31 practitioners focused on investigating the relationship between three main activities in safety engineering, and an agile approach (safety requirements development, hazard analysis, and safety case development) is presented in the study of Doss and Kelly [19]. 80% of the practitioners with experience in SCS development and agile development methods and from different industrial sectors either agreed or strongly agreed that agile methods could be integrated with safety-critical systems development.

Since requirements-related errors have been reported as the main cause of accidents [22], it is paramount to have a safe requirements engineering process [20]. Requirements engineering is a software engineering area that addresses the elicitation, documentation, and management of system requirements. According to Kotonya and Sommerville [1], it comprises five main activities:

- **Requirements Elicitation:** it is related to requirements gathering and discovering the stakeholders' needs. In this activity, the system requirements are defined by consulting various domain knowledge sources, which may include customers, similar systems, standards to be compliant, and business artifacts, for example. Several techniques can be used to elicit requirements, such as interviews, ethnography, brainstorming, prototyping, etc. Kotonya and Sommerville [1] define four main tasks to be executed at this stage: Establish objectives, Understanding background, Organise knowledge, and Collect requirements.

- **Requirements Analysis and Negotiation:** the primary goal of this activity is to reach agreements about the system to be developed to satisfy the stakeholders. The tasks performed involve the analysis, prioritization, and negotiation of requirements.
- **Requirements Specification:** the set of negotiated requirements obtained in the previous activity is documented at an adequate level for stakeholders at this activity. To specify the requirements can be used natural language, formal models, and diagrams such as the ones of the Unified Modeling Language (UML) [26] and goal-oriented modeling languages [27][28].
- **Requirements Validation:** the requirements document is verified against consistency, completeness, and accuracy. The outputs are a list of reported problems and a list of actions in response to those problems.
- **Requirements Management:** is an activity conducted simultaneously with the others previously described that handles changes in the requirements and their traceability.

A comprehensive study about the approaches that have been proposed to elicit, model, specify, and validate safety requirements in the context of SCS was developed by Martins and Gorschek [21]. Their work also analyses to what extent such approaches have been validated in industrial settings. Since the focus of this chapter is agile RE, in the next section, we describe the main features of agile methods.

4.2 Agile Methods

The software development paradigm usually followed to develop an SCS is the traditional one [2][25] that requires detailed documentation to ensure system safety and reliability. It is also called a plan-driven approach since all process activities are planned beforehand [25], and the progress is evaluated concerning the plan [2]. Accordingly, it is necessary to specify the requirements and design, build, and test the system [2].

On the other hand, the agile methods rely on building new releases of software as a series of increments in which their development depends on the team progress and customer priorities [2]. The principles that guide the agile paradigm of managing projects were established in the release of the agile manifesto in 2001 [3] signed by 17 professionals that claim:

> *Individuals and interactions* over processes and tools.
> *Working software* over comprehensive documentation.

> *Customer collaboration* over contract negotiation.
> *Responding to change* over following a plan.

According to the manifesto, the items on the left of the sentences have more value in this paradigm. The agile methods are based on incremental system delivery and share some features, as explained by Sommerville [2]:

- Customers should participate actively in the development process;
- The processes of specification, design, and implementation are interweaved;
- The system is developed in a series of incremental versions where the customer determines the requirements to be implemented in each version;
- Requirements and system design can change frequently.
- User interfaces are frequently elaborated using an interactive development system.

A comparative analysis among agile software development methodologies can be found in the work of Matharu et al. [18]. The authors concluded that there is a higher adoption of Scrum in relation to Extreme Programming and Kanban methodologies. In the following sections, we briefly define these most adopted methodologies: Scrum and XP.

4.2.1 Scrum

Scrum [5] relies on the concept of a sprint cycle that is a planning unit that corresponds to an iteration where requirements are prioritized, and the software is developed [2]. When a sprint ends, a small version of the system is delivered. According to Sommerville [2], the main features of this agile process are:

- **Sprints have a predefined length**: It usually lasts 2-4 weeks.
- **A product backlog contains the list of work to be done on the project**: Such backlog is constantly reviewed, and priorities and risks are assigned with the customer's active participation.
- **The requirements selection is a collective task**: the team that is close to the customer selects the functionalities to be developed during each sprint.
- **Daily meetings**: they are conducted with all team members to assess the progress and, if necessary, reprioritize work.
- **Scrum master**: it is a role in which he/she is responsible for all communications with the customer.

- **Progress review**: when a sprint ends, the work is reviewed and presented to stakeholders.

4.2.2 XP

The eXtreme Programming (XP) is an example of an agile software development process. It is based on 13 principles or practices [6] explained below.

1. **Planning game**: The system's functionalities are divided into stories that are implemented in iterations according to customer priorities.
2. **Small releases**: A small, but working, version of the system is built in each iteration and available to the customer.
3. **Metaphor**. A metaphor or set of metaphors shared between the customer and programmers defines the system shape.
4. **Simple design**: The design is kept simple with only the necessary classes and methods, and the tests are conducted continuously.
5. **Tests**: Programmers and customers, respectively, design unit tests and functional tests. The system is frequently tested, and it should run correctly.
6. **Refactoring**: Changes are performed in the system design to improve it, but all tests should continue running.
7. **Pair programming**: Two people develop the code at one screen, keyboard, and mouse.
8. **Continuous integration**: New code is combined with the current system frequently without breaking the tests already performed or the changes are rejected.
9. **Collective ownership**: The code belongs to the team, and any developer can improve it whenever he/she sees an opportunity.
10. **On-site customer**: A customer is available full-time at the company to solve the doubts of the team.
11. **40-hour weeks**: The team should work on the predefined schedule. When overtime occurs frequently is a sign of deeper problems in the team.
12. **Open workspace**: it is recommended that the team be allocated in a large room with small cubicles, and the center is designed to pair programmers.
13. **Just rules**: A set of rules, that can be changed according to necessity, is followed by the team.

Although agile methods may be hard to adopt in complex projects such as SCS or large organizations [2][7], they are obtaining great notoriety in the last few years. They have offered a collection of practices that allow sustainable gains in productivity and quality in software development [4]. In this context, it is possible to integrate agile method practices in the RE of SCS [4][23][24].

In the next section, we discuss the RE stage involved in developing a safety-critical system (SCS) when an agile development process is followed.

4.3 Agile Requirements Engineering in SCS

The interest of agile practices for requirements engineering has been studied by several works [17][23]. An investigation of how RE is performed in projects that follows agile methodologies is presented in the work of Medeiros et al. [17]. They examined which RE techniques to elicit and to specify requirements are being used in projects that adopt agile methodologies. Their results are explained in Sections 3.1 (elicitation) and 3.3 (specification).

Cao and Ramesh [23] discuss a qualitative study with 16 organizations' participation to analyze the RE practices agile and their benefits and challenges. They observed that the companies performed seven agile RE practices: face-to-face communication over written specifications, iterative requirements engineering, requirement prioritization goes extreme, managing requirements change through constant planning, prototyping, test-driven development (TDD), and use review meetings and acceptance tests.

In the following sections, we present approaches according to the main phases of requirements engineering [1].

4.3.1 Requirements Elicitation

In requirements engineering, the elicitation corresponds to the activities required to discover system requirements. In the study of Medeiros et al. [17], seven strategies are being used to elicit requirements in agile projects: interviews (the most used), JAD, focal group, brainstorm, questionnaire, trawling, and workshop.

In the construction of an SCS, it is necessary to conduct safety requirements elicitation, hazard analysis, and risk analysis to define the hazards, safety requirements, and their risk. These activities can be performed iteratively and incrementally, according to the survey of Doss and Kelly [19], and need to be reassessed during software development. Accordingly, in the literature, there are studies [8] that aim to cover this gap.

A framework that combines safety activities and safety risk assessment techniques (SRATs) into requirements elicitation is proposed by Yeow and Kia Chiam [8]. The Safety Risk Assessment Techniques in Requirements Elicitation (SaTRE) Integration Framework splits the requirements elicitation into three stages:

1. Pre-elicitation: its goal is to identify the project objectives, scope, system boundaries, and relevant requirements sources.
2. Midst of elicitation: it is performed to discover existing requirements and the new requirements.
3. Post-elicitation: is conducted after applicable stakeholders approved that there are no additional requirements or scenarios to be added.

The SaTRE Framework relies on a repository to store the information about SRATs, guidelines, and templates to help the requirements and safety engineering teams to select and use the techniques. A web-based tool [9] was developed to support the application of this framework.

4.3.2 Requirements Analysis and Negotiation

The phase of analysis and negotiation comprehends the prioritization and selection of the requirements to be implemented after reaching a consensus with the customer. The prioritization is conducted in agile processes considering the business value, and it is performed at the beginning of each software development cycle [23].

Requirements prioritization was recognized as necessary by 77% of the practitioners that participated in the survey described in [19]. Agile approaches address the volatility of requirements by developing the system in cycles with a small and predefined set of requirements.

Stallhane, Myklebust, and Hanssen [13] proposed changes in the Scrum method to apply it in the development of safety-critical systems. The Safe Scrum process defines two product backlogs: the functional product backlog, commonly adopted in Scrum projects, and one safety product backlog, employed to handle safety requirements. The safe Scrum process, which contributes to the requirements analysis and negotiation tasks, has already been applied in real-life projects [13].

A requirements analysis method that relies on viewpoints and incorporates safety analysis as part of requirements analysis is explained in the work of Kotonya and Sommerville [20]. The views considered in the method are the direct ones that consider the customers that use the system functionalities, and some indirect ones where visions of stakeholders that are interested in some or all functionalities but do not interact directly.

4.3.3 Requirements Specification

Specification of requirements in the context of an SCS is a challenging activity [12] since it is necessary to elaborate a correct, complete, and unambiguous document [11] to avoid misinterpretations that could lead to accidents and safety incidents. In this context, the agile principles help to reduce such challenges through short iterations, quick feedback, and active stakeholders [11].

In the survey of Medeiros et al. [17], user stories and wireframes are the most used techniques to specify requirements in agile projects among the 21 identified. Other techniques reported were: XXM, Activity Diagram, AUC, ALC, ACC, Mind Map, INVEST, and GPM, and Cucumber.

Some approaches are available in the literature to solve the requirements specification issues in an SCS [4][11]. An Agile Requirements Specification Process for Regulated Environments (ARES) is proposed by Marques and Cunha [4] aiming to ensure compliance, in the perspective of requirements specification, with six safety-critical standards: RTCA DO-178C, IEC 62304:2015, ECSS-E-ST-40C, IEC 61508-3, ISO/IEC/IEEE 12207, and IAEA SSG-39.

ARES adopts Scrum-like sprints proposed to develop the software specifications where at each sprint, a set of prioritized System Requirements (Sprint Backlog) is required as input and produces four artifacts as output: software user stories, the High-Level Architecture (HLA) that defines the main interface between software product components, test cases, and traceability information [4].

The process defines four methods with 13 activities [4]:

- **M1: Analysis, Selection, and Prioritization of System Requirements**: dedicated to analyze and identify system requirements applied to software and prioritizing system requirements for sprints.
- **M2: Definition of Software Specifications:** it generates software user stories, HLA, test cases, and corrections are conducted.
- **M3: Revision of Software Specifications:** in which software user stories are traced, software specifications are reviewed, burndown chart is updated, and problems are reported.

- **M4: Execution of the Sprint Retrospective**: it defines three tasks: close problems, analyze and confirm issues, and schedule sprint for problem correction.

The solution for specifying agile artifacts more accurately, accordingly to Leite [11], is an Agile Safety Process, whose aim is to select which artifacts or parts thereof are necessary to document failure detection and containment, as well as procedures for leading the system to a safe state that includes traceability requirements, standard-related constraints, and domain experts. The generated artifacts are input to a semi-automated methodology proposed to the specification of agile artifacts that considers safety aspects and adopts user stories to specify the requirements.

4.3.4 Requirements Validation

System validation is a phase where the requirements previously elicited are validated and formally agreed upon by the customer. Accordingly, it is evaluated whether the right system is being developed. The most common technique used in this phase is meetings with the customer, and some approaches rely on this technique [14]. Other practices used to validate requirements in agile projects are Test-Driven Development (TDD), acceptance tests, and test cases [17].

The Safe Scrum process [14], which proposes developing the system in small versions, may require some adaptations in the software lifecycle depending on the safety standard followed. Stallhane, Myklebust, and Hanssen [13] discuss some changes and adaptations necessary to use the process along with IEC 61508-3, IEC 60880, and EN 50128 in order to turn the process acceptable to Scrum and the safety evaluators [14]. When the first standard is followed, how to construct a validation plan and test and check the outputs from software safety lifecycle activities should be changed and the definition of requirements for software module testing. The implementation verification and test requirements and component testing should be adapted to the second and third standards.

When Safe Scrum is adopted, a validation of the Reliability, Availability, Maintenanceand Safety (RAMS) is performed after the sprints are finished. According to Stallhane, Myklebust, and Hanssen [13], the final RAMS validation is expected to be less extensive compared to other development paradigms, considering that most of the system has been incrementally validated during the sprints. This incremental validation contributes to reducing the time and cost needed for certification.

4.3.5 Requirements Management

Certification is mandatory in several safety-critical domains such as avionics, medical, and automotive. Besides being a compulsory activity by some standards, such as DO-178B, requirements traceability (or management) contributes to deal properly with changes during system development [10] since any change in an SCS usually involves repeating the certification process.

According to Cao and Ramesh [23], two types of requirements change: adding or dropping features and changing already implemented features. They also discussed that changes are easier to implement and cost less in agile development since fixed plans are not performed, and the planning is revisited as soon as the change occurs.

Considering the need for certification where the authorities review the produced artifacts, it is necessary to maintain them appropriately along the system lifecycle. Furthermore, requirements traceability helps enhance the change management process and to ensures the system's correctness and quality [15].

An agile modeling approach to managing requirements traceability is presented in the work of Taromirad and Paige [10]. It is based on a domain-specific requirements traceability approach that defines a Domain-Specific Modelling Language (DSML) to incrementally construct a traceability structure for a specific domain or project. The language is designed for the current project, target domain, and traceability goals and the traceability metamodel is built incrementally through iterations.

In Taromirad and Paige [10] approach, each iteration ends when a change is detected either because there is a requirement lacking or a new one arises and, consequently, a new iteration begins. This cycle is repeated over the project, and the traceability scheme will be extended whenever needed. The approach is exemplified in the context of the SCS domain, in which the traceability model is linked to a safety case model.

Traceability is also a benefit of Leite's solution [11]. Besides providing an iterative solution to specify system requirements in an SCS, the solution also contributes to offer evidence to authorities that safety actions have been taken and that there is end-to-end traceability of this set of agile artifacts.

The Safe Scrum process [14] that uses two product backlogs, as discussed in Section 3.2, links the functional requirements of the functional product backlog to the safety requirements in the safety product backlog that are affected by which functional requirements. This relationship, which could be maintained by using simple cross-references in the two backlogs, is a prerequisite to understanding how the requirements are connected, making them easier to manage and to perform impact analysis.

For more information about requirements traceability, a Systematic Literature Review (SLR) is presented in the work of Tufail et al. [15] that selected 33 studies from 2010 to 2017 about requirements traceability. The SLR compared seven models and 14 tools, discussed ten challenges, and inferred that DOORS and Traceability Meta Model are the best requirement traceability tool and model, respectively. Trindade and Lucena [16] present a survey regarding challenges related to traceability in agile environments.

4.4 Traditional x Agile Requirements Engineering

A comparison between traditional software engineering and agile software development can be found in several works [2][29]. In this chapter, we are concerned with agile requirements engineering in SCS and, because it is an emerging field, we observe a tendency in the approaches to balance the practices of traditional or conservative methodologies with the agile ones [4][24][25]. According to Heeager and Nielsen [24], companies should pay attention to four areas when combining traditional and agile processes with building an SCS: requirements, documentation, life cycle, and testing.

In this chapter, our concern is the first two areas. Regarding requirements, the main issues and possible solutions in combining these processes are:

- **Achieving flexible requirements and maintaining traceability**: to conciliate the processes, Heeager and Nielsen propose detaching functional and safety requirements at the beginning of the development and complementing formal requirements with user stories.
- **Documenting the requirements in user stories**: this issue can be addressed by adapting user stories to include risks and write safety stories

The challenges and solutions related to the use of documentation comprehend:

- **Delivering the suitable quantity and the proper type of documentation for demonstrating software safety:** among the solutions to mesh the traditional and agile processes are handling documentation as part of the product demanded by the customer and producing only documentation required according to its purpose;

- **Addressing change management and maintaining traceability:** means handling these challenges involves developing the documentation iteratively before implementation, managing the documentation as an isolated activity during each iteration, and defining sub-teams in charge of documentation. These practices are considered in the ARES method [4] discussed in Section 3.3. Several artifacts are generated aiming to document the SCS and obtain evidence for certification, but they are constructed iteratively as suggested by the agile methodologies.

Although meshing traditional and agile methodologies can be achieved [24], it is observed differences among them in several phases of software development:

- **Requirement prioritization**: it is an extensively adopted agile practice, but this activity is conducted with some differences considering its methodology [23]. In their survey, Cao and Ramesh [23] concluded that in traditional RE, requirements are typically prioritized once. However, in the 16 companies they studied, the requirements are prioritized during the planning meetings at the beginning of each agile development cycle. In this regard, another significant difference is the factors that drive prioritization. While there are many factors, such as business value, risks, cost, and implementation dependencies, in the traditional methodologies, the main factor considered in agile processes is the business value.
- **Requirements validation:** more emphasis is dedicated to this area in agile RE practices. Nevertheless, formal verification is usually not performed in agile development since formal modeling of detailed requirements does not exist [23]. Instead, consistency checking or formal inspections are conducted.
- **Cost of requirements management:** it is cheaper to accommodate changes in agile software development because the planning is not fixed, and requirements frequently change on the contrary of traditional methodologies [23]. Another difference is the factor that leads to these changes, as discussed in Section 3.5.
- **Contractual changes:** in conventional methodologies, the requirements should be defined when the project starts. It is necessary to decide all functionalities of the system and how they will be combined with the hardware [25]. In contrast, agile approaches rely on intensive customer participation to refine and prioritize requirements at the beginning of each sprint.

4.5 Case Studies

The development of a safety-critical system following an agile process has little empirical evidence of its effectiveness. Nevertheless, some case studies [24][25] are available in the literature, and they are discussed in the following sections.

4.5.1 Pharmaceutical Company

A pharmaceutical company is the case study in which a Scrum process was applied to combine agile software development in a traditional safety-critical project in [24]. The product should be compliant with several safety standards, including those associated with the US FDA, and started with more than 100 managers, engineers, and developers involved.

Regarding the RE process adopted, the requirements for the medical device were mostly defined based on a user survey conducted at the beginning of the project. Then, hazard analysis was performed, resulting in the generation of requirements that were frozen and written into several documents, such as the device specification and the hazard analysis.

The agile practices adopted included two weeks sprints involving:

- unit-testing with 100% condition-decision coverage;
- internal delivery of the increments;
- software backlog that consisted of the software specification;
- post-its to write the tasks to be implemented;
- burndown charts to trace the progress of the sprint;
- daily stand-up meetings;
- peer-review of tasks;
- a formal report of errors;
- assessment of the process and the current increment at the end of each sprint.

Among the challenges faced by the team, the authors mention the underestimation of the complexity of the tasks leading to the difficulty to define their size. Regarding RE, the issues were related to requirements uncertainty and requirements changeability. The former was a result of poor requirements specifications and because no customer was present; the latter is associated with frequent requirements changes.

4.5.2 Avionics Company

Storer and Islam [25] present a case study where semi-structured interviews were performed with software engineers employed in a large avionics company in the United Kingdom. Their goal was to investigate the company's experiences in the adoption of agile software development to SCS and the difficulties experienced.

They observed that the studied company also follows Scrum, and practices implemented by the company were:

- daily meeting;
- sprints of one or two weeks long;
- a scrum master responsible for organizing the activities;
- the team created a plan at the start of each sprint, using a Jira or Kanban issue board to track progress.
- the scrum master estimates the effort in terms of story points in the sprint based on team size and availability;
- the teams do a "T-shirt size" estimation of the tasks and record this on Jira boards;
- Microsoft Project is used for long term planning;
- tasks are selected from the backlog and included in sprints.

Finally, fourteen issues in adopting agile methods and practices in the context of software development for safety-critical systems are discussed in the case study. Four themes group the issues: pressure for waterfall; coordination amongst stakeholders; documentation demands, and cultural challenges.

4.6 Conclusions

The development of safety-critical systems involves producing several artifacts to document the system, communicate requirements among stakeholders, and conduct hazard and risk analysis to demonstrate that the system is safe. All this information is stored and should be maintained through the system lifecycle.

Software development processes can be followed to guide this development by defining phases and providing practices to be followed. Such processes can be classified as traditional or conservative, agile or emergent software lifecycles. Traditional software methodologies like a waterfall are usually followed to build an SCS. However, the adoption of agile processes is increasing even though it is used in a mixed approach in the projects.

In this chapter, we discussed the main characteristics of agile software development methodologies like Scrum and XP. We also defined the main phases of requirements engineering and agile approaches used in each of these phases, and we compared the traditional versus agile requirements engineering. Finally, two case studies were presented.

Acknowledgments

We would like to thank Universidade Federal de Pernambuco (UFPE).

References

[1] G. Kotonya, I. Sommerville. Requirements engineering: processes and techniques. John Wiley & Sons, Inc., 1998.
[2] I. Sommerville. Software engineering 9th Edition. 2011.
[3] M. Fowler, J. Highsmith, J. The agile manifesto. Software Development, 9(8), 28-35, 2001.
[4] J. Marques, A. M. da Cunha. ARES: An Agile Requirements Specification Process for Regulated Environments. International Journal of Software Engineering and Knowledge Engineering, v. 29, n. 10, p. 1403-1438, 2019.
[5] K. Schwaber and M. Beedle, Agile Software Development with Scrum. Prentice Hall, Upper Saddle River, 2001.
[6] K. Beck and C. Andres, Extreme Programming Explained: Embrace Change. Addison-Wesley, Boston, 2004.
[7] T. Dyba and T. Dingsoyr, What do we know about agile software development?, IEEE Software. 26, pp. 6–9, 2009.
[8] E. Yeow, C. Yin Kia. Integration of Safety Risk Assessment Techniques into Requirement Elicitation. In: SoMeT, pp. 256-270, 2014.
[9] E. Yeow, C. Yin Kia. A web-based system for integrating safety techniques into requirements elicitation. In: 2015 9th Malaysian Software Engineering Conference (MySEC), pp. 87-92, 2015.
[10] M. Taromirad, R. F. Paige. Agile requirements traceability using domain-specific modelling languages. In: Proceedings of the 2012 Extreme Modeling Workshop. 2012. p. 45-50.
[11] A. I. M. Leite. An Approach to Support the Specification of Agile Artifacts in the Development of Safety-Critical Systems. In: 2017 IEEE 25th International Requirements Engineering Conference (RE), pp. 526-531, 2017.

[12] J. Hatcliff, A. Wassyng, T. Kelly, C. Comar, P. Jones. Certifiably safe software-dependent systems: challenges and directions. In: Future of Software Engineering Proceedings, pp. 182-200, 2014.

[13] T. Stålhane, T. Myklebust, G. Hanssen. Safety standards and Scrum–A synopsis of three standards. In Nbl. SintefNo.

[14] T. Stålhane, T. Myklebust, G. Hanssen. The application of safe scrum to IEC 61508 certifiable software. In: 11th Int. Probabilistic Saf. Assess. Manag. Conf. Annu. Eur. Saf. Reliab. Conf. 2012, PSAM11 ESREL 2012, vol. 8, pp. 6052–6061, 2012.

[15] H. Tufail, M. Masood, B. Zeb, F. Azam, M. Anwar. A systematic review of requirement traceability techniques and tools. In: 2017 2nd International Conference on System Reliability and Safety (ICSRS), pp. 450-454, 2017.

[16] G. O. Trindade, M. Lucena. Requirements Traceability in Agile Methodologies: A Exploratory Survey. In: Proceedings of the XII Brazilian Symposium on Information Systems on Brazilian Symposium on Information Systems: Information Systems in the Cloud Computing Era-Volume 1, pp. 478-485, 2016.

[17] J. Medeiros, D. C. Alves, A. Vasconcelos, C. Silva, E. Wanderley. Requirements Engineering in Agile Projects: A Systematic Mapping based in Evidences of Industry. In: CibSE, pp. 460, 2015.

[18] G. S. Matharu, A. Mishra, H. Singh, P. Upadhyay. Empirical study of agile software development methodologies: A comparative analysis. In: ACM SIGSOFT Software Engineering Notes, v. 40, n. 1, p. 1-6, 2015.

[19] O. Doss, T. P. Kelly. Challenges and opportunities in agile development in safety critical systems: A survey. ACM SIGSOFT Software Engineering Notes, v. 41, n. 2, pp. 30-31, 2016.

[20] G. Kotonya, I. Sommerville. Integrating safety analysis and requirements engineering. In: Proceedings of Joint 4th International Computer Science Conference and 4th Asia Pacific Software Engineering Conference, pp. 259-271, 1997.

[21] L. Martins, T. Gorschek. Requirements engineering for safety-critical systems: A systematic literature review. Information and software technology, v. 75, p. 71-89, 2016.

[22] N. Leveson. Engineering a safer world: Systems thinking applied to safety. The MIT Press, 2016.

[23] L. Cao, B. Ramesh. Agile requirements engineering practices: An empirical study. IEEE software, v. 25, n. 1, pp. 60-67, 2008.

[24] L. T. Heeager, P. A. Nielsen. Meshing agile and plan-driven development in safety-critical software: a case study. Empirical Software Engineering, pp. 1-28, 2020.

[25] G. Islam, T. Storer. A case study of agile software development for safety-Critical systems projects. Reliability Engineering & System Safety, 2020.

[26] J. Rumbaugh, I. Jacobson, G. Booch, Grady. The unified modeling language. Reference manual, 1999.

[27] E. Yu, J. Mylopoulos. Why goal-oriented requirements engineering. In: Proceedings of the 4th International Workshop on Requirements Engineering: Foundations of Software Quality, pp. 15-22, 1998

[28] J. Vilela, J. Castro, L. E. G. Martins, T. Gorschek, T., & Silva, C. Specifying safety requirements with gore languages. In: Proceedings of the 31st Brazilian Symposium on Software Engineering, pp. 154-163, 2017.

[29] K. Beck. Embracing change with extreme programming. Computer, v. 32, n. 10, p. 70-77, 1999.

5

A Comparative Study of Requirements-Based Testing Approaches

J. Santos and L. E. G. Martins

Federal University of São Paulo, Institute of Science and Technology, São José dos Campos, Brazil
E-mail: jemison321@gmail.com; legmartins@unifesp.br

Abstract

Context: Identifying a methodology for testing the requirements of a software system can be a daunting task. Several requirements testing approaches can be found in the literature, but for critical safety systems, the engineer needs to focus more on requirements and their testing. Objective: This selected article aims to perform two safety requirements testing approaches based on a systematic literature review. The objective is to compare the methodology's application and verify the method's performance, considering the variables of application time in models and test cases. Method: An experiment was performed that applied two approaches to software requirements testing, namely: activity diagram based requirements testing and behavior tree based requirements testing. Results: The study pointed out inconsistencies in requirements and showed that in this context of the application, the behavior trees had a shorter development time than activity diagrams. The time taken to generate test case specifications was shorter compared to the generation of test behavior trees. Conclusion: The application of two model-based requirements testing approaches has helped to identify several features that bring advantages and disadvantages for the software engineer, as well as provide improvements in the requirements specification.

Keywords: Requirements Models, UML, BML Student's t-test.

5.1 Introduction

Software requirements, when written in natural language, can convey ambiguous information due to linguistic variations. Engineer's misinterpretation of requirements can create problems for the customer who will receive a product with different features than requested. A software system, whether for use in everyday activities or data processing for advanced users, will always be able to have new functionality incremented over time so that new requirements will be elicited. As new modules and new features will be implemented, the size and complexity of the system will increase. [5, 9, 16].

The software requirements, once defined, will be modeled to better understand the system entirely in its logical structures. It becomes easier to find what is relevant to the system when requirements are modeled [1, 8]. A testable requirement model represents a set of software requirements, which contains specific features to simplify the understanding of the view of system functionality and to assist in generating test cases [3, 15].

It is noted that software requirements testing can be performed in several ways so that the specification must have testable requirements and meet characteristics such as completeness and unambiguity in the test context. This is a basic premise for generating testable requirements models [8, 18].

Based on a systematic literature review by Santos et al. [14], the main approaches used to perform the software requirements test were raised. This research served as the basis for the choice of two approaches that obtained high scores in the quality assessment. They were: Requirements-Driven Testing with Behavior Trees and Model-Based Testing of System Requirements using UML Use Case Models.

This chapter aims to discuss which of the approaches selected in the systematic literature review [14] is best suited for testable modeling safety requirements in the context of an insulin infusion pump. The chapter is organized as follows: Section II presents background and related work; section III shows how the experiment was performed and which metrics were defined for its analysis; section IV presents the results and discussions of the results obtained; section V shows the conclusions of the chapter, and section VI shows future work.

5.2 Background and Related Work

In this section, we present contextual articles similar to the experiment performed. They were found through a systematic literature review and are described here to demonstrate the state of the art obtained during the research. Thus, the next subsections present these works and detail the approaches chosen by the review for application in the experiment.

A. *Systematic Literature Review Papers*
We found several articles by authors who presented approaches to perform requirements-based testing. The Legeard et al. [10] approach explains that the Model-Based Testing (MBT) technique is spreading, and each year it is possible to find more research being developed in the testing area based on requirement models. This technique has contributed to the automation of test generation and execution based on software requirements models.

Legeard et al. [10] researched the MBT application and concluded that it is not able to solve all software problems. However, MBT is an analytical technique that promotes success in the state of practice of functional testing, requirements testing, and increases productivity with improved functional coverage of the requirement. Hammer et al. [7] performed the use of automated tools to track requirements and their test cases. In his study, Hammer et al. [7] explain that according to the criticality of software systems, implementation engineers have some difficulties in specifying requirements in order to be clear to analysts and developers.

Hammer et al. [7] further discuss that there are tools that help with requirements management and open doors for using useful metrics for characterizing and assessing software requirements risk. Hammer et al. [7] focus on which metrics are essential for providing benefits that are associated with early detection and correction of requirements issues. The main objective of his work is the application of metrics through a CASE requirements management tool.

An article published by Gutiérrez et al. [6] presents a model-based testing approach with a focus on functional testing. They are seeking to support end-system requirements as previously defined by the user. By drawing four types of metamodels, the author points out that not all aspects relevant to a functional requirement were included. Aspects such as the Navigational Development Techniques that define models through the UML and another approach that makes the generation of test cases through the requirements models and the colloquial language were addressed.

Another work that also applies UML for test case generation is presented by Cartaxo et al. [2]. This work shows a procedure in which functional test cases are systematically generated. The procedure presented was modeled according to the test patterns based on sequence diagram models and translated to Labeled Transition Systems (LTS). The results were presented through a case study.

B. *Selected Approaches to Study*

According to the results obtained from a systematic literature review, we performed [14], two articles with different views on the generation of test cases were listed, using different approaches.

Hasling et al. [8] study describes the author's experience in applying UML use cases as a basis for generating test cases in a Siemens Medical project. They argue that domain experts are more likely to interpret case diagrams than test specs. They also claim that the use of case models corroborates with verification and validation, as these two components can be met, and in addition to enhancing software requirements through the model, which can also improve test coverage. Traceability of requirements is more natural, says Hasling et al. [8]. The association of requirements with models makes the reverse process also better.

Using the TDE/UML tool developed by Siemens Corporate Research, Hasling et al. [8] generated test cases based on the software's UML models. By applying a path generation algorithm, the system seeks to find activity diagram variations that are compatible with the variables expressed by the tester linked to an equivalence class, and unviable paths are discarded.

Hasling et al. [8] presented errors that were found in models that led to problems with test case generation, such as Use-free cases, illegal activity states, and illegal restrictions. Finally, Hasling et al. [8] conclude by explaining what the focus of the work was to prioritize testing software requirements rather than the functionality of user interface behavior.

According to Wendland et al. [18], the article presents a methodology for generating test requirements that are based on behavior engineering. The objective was to apply behavior tree diagrams and extend them with testing activities. The authors explain that requirements described in natural language are hardly unambiguous, and ease of understanding requirements will influence the effectiveness and efficiency of validations and verifications.

Wendland et al. [18] are motivated by the challenges posed by model-based requirements testing methodologies. Working at the first stage of software development, the authors explain, behavioral engineering can capture and validate requirements systematically. Complex specifications that would be handled by the development team are handled through a scalable and repetitive methodology.

The composition of behavioral engineering models is divided into two parts: Behavior Modeling Language (BML) in conjunction with Behavior Modeling Processes (BMP). Wendland et al. [18] further state that due to the ease of readability of the generated models, stakeholders can check the requirements several times throughout the process. By proposing the so-called test augmentation, Wendland et al. [18] introduced a new behavior tree model, and this model provided a traceability status that, together with the information described in the model was considered sufficient to perform system requirements testing.

With the results obtained, Wendland et al. [18] listed advantages such as improved test requirements elicitation process with the view of a tester - with regard to the requirements in the test process, the information provided helps in testing the requirements specification and need not be gradually determined; The augmentation test provided greater systematization for defining the requirement to be tested with rules applied in the modeling as a whole and maintaining the standards applied by the IEEE830 standard.

5.3 Experiment Design

In this section, we present the protocol of the experiment that we have conducted. The protocol shows how two approaches will be compared based on an experiment. The two test approaches are model-based, focusing on the elaboration of requirements models that will assist in the generation of test cases.

A. *Scope*
To experiment, we sought to answer the following research questions:

RQ1: Which approach is faster to generate test cases from requirement models?

RQ2: Which approach can generate more test cases from requirement models?

RQ3: In which approach is the highest number of models based on the requirements?

The objectives of the experiment are displayed in Table 5.1.

B. *Planning*
The design of the experiment was defined in six phases: Context Selection, Variable Selection, Hypothesis Formulation, Subject Selection,

Table 5.1   The objective of the experiment.

Analyze	Ease of use and manual modeling. which Two model-based software safety testing approaches are used: UML [8] use cases and activity diagram requirements modeling and BML behavior tree modeling [18].
For	Comparison.
With respect to	Ease of use and effectiveness in test case generation of use case and behavior tree based approaches.
With the point of view of	Researcher.
In the context of	development of control software for a low-cost insulin infusion pump.

Experiment Design, Instrumentation Evaluation, and Validity. These phases are described in the next subsections.

1) Context Selection: The context of the experiment is a low-cost insulin infusion pump, developed by an interdisciplinary research group from UNIFESP, in partnership with a company located in the city of São José dos Campos.

The insulin pump hardware consists of a non-rechargeable power supply, a conventional syringe as an insulin reservoir, a stepper motor for the infusion process that connects to the mechanical system to move the driver-controlled syringe plunger, and provides the information to the user through a GLCD display.

Figure 5.1 shows the components of the insulin pump, shown in figure (a) the infusion pump prototype, in figure (b) the infusion pump housing, and separately the insulin reservoir, and figure (c) the hardware.

The main requirements of the software are:

- Information display: The pump should display on the GLCD display its approximate battery levels, insulin amount, time, fault alarm, and description of the engine's usage status.
- Check Numeric Data on Battery and Insulin Levels: The user has the option of checking battery and insulin status on a separate screen for more accurate analysis.
- Set up and enable infusion profiles: The user can configure five different basal infusion profiles (Insulin concentration that the patient should receive continuously throughout the day) every 24 hours.
- At any time, the user can reset the stored profiles.

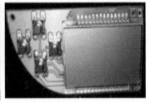

(a) Prototype
Overview

(b) Insulin Case and
Reservoir

(c) Electronic scheme
with control buttons
and LCD

Figure 5.1 Insulin pump components [12].

- Set and enable rapid infusion (Bolus infusion – Additional insulin dose delivered faster than basal.): The user can configure the amount of insulin they currently want to infuse.
- Enable insulin reservoir exchange: The user will have a step-by-step explanation on the screen of how to change the insulin reservoir.
- Critical Alerts. For each problem, such as a low battery or a lack of insulin, the pumps should use audible and visible alerts.

Figure 5.2 shows the pump context diagram presented by Martins et al. [12] in the requirements specification document. This template contains all hardware components that are monitored or controlled by the software of the insulin infusion pump.

The experiment is performed manually and focuses on applying two model-based requirements testing approaches and the time-cost to generate requirement models. This experience addresses a real problem, namely the modeling of insulin pump requirements to assist in generating test cases. Comparing approaches can motivate other software engineers to adopt model-based approaches to creating testable requirement models and contribute to the strengthening of the model-based requirements testing paradigm.

2) Variable Selection: Independent variable: Software requirement model. Dependent variables: a set of test cases and test behavior trees generated from requirements modeling, as well as the time-measured effort to generate all models, cases, and test trees.

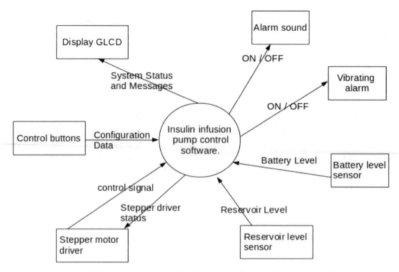

Figure 5.2   Context model of the insulin infusion pump [12].

3) Hypothesis Formulation: Formulating hypotheses in an experiment is an important task to describe what will be analyzed within the defined context. In this experiment, we modeled diagrams according to the UML language and behavior trees of the BML language. Subsequently, the generated requirements models were used to create test cases and test behavior trees. One method is believed to be more effective than another.

1) **The null hypothesis,** H_{01}: There is no difference in time cost for modeling requirements in activity diagrams (AD) and behavior trees (BT).
   H_{01}: Cost_Time(AD) = Cost_Time(BT)
   Alternative hypothesis H_{11}: Cost_Time(AD) $\neq$ Cost_Time(BT), AD $<$ BT or AD $>$ BT;

2) **The null hypothesis,** H_{02}: There is no difference in time cost for generating test cases created with the activity diagram (TC) and behavior tree (TBT) requirement models.
   H_{02}: Cost_Time(TC) = Cost_Time(TBT)
   Alternative hypothesis: H_{12} : Cost_Time(TC) $\neq$ Cost_Time(TBT), TC $<$ TBT or TC $>$ TBT;

3) **The null hypothesis,** H_{03}: The same number of test cases can be generated from behavior tree-based requirement models (TBT) and activity diagrams (TC).

H_{03}: Amount(TBT) = Amount(TC)

Alternative hypothesis, H_{13} : Cost_Time(TBT) $\neq$ Cost_Time(TC), TBT < TC or TBT > TC;

4) **The null hypothesis,** H_{04}: The same number of requirements are modeled in activity diagrams (RAD) and behavior trees (RBT).

H_{04}: Amount(RAD) = Amount(RBT)

Alternative hypothesis, H_{14}: Amount(RAD) $\neq$ Amount(RBT), RAD < RBT ou RAD > RBT.

4) Subject Selection: The experiment was conducted by a master student who performed the whole process of generating requirements models, test cases, and evaluation of results. The experiment was performed on a software system of an insulin infusion pump.

5) Experiment Design: A comparison of the methods used in the model-based software testing approaches of Hasling et al. [8] and Wendland et al. [18] was performed to verify in both approaches the characteristics and metrics as follows:

- Model Completion: Verify that the model describes the requirement in detail;
- Ease of understanding of the model: Check particularities that make the understanding of the model easier to comprehend;
- Traceability: Discuss the possibility of tracking and describing the life of the requirement in both directions;
- Modeling ease: Verify the cost in time for requirements modeling according to each approach;
- Number of test cases: Check the number of test cases each model produces;
- Number of models: Check the number of models generated from the requirements;

Software requirements models of a low-cost insulin pump will be generated, seeking the strengths and weaknesses of each approach; the described metrics were created and will be analyzed by the researcher.

6) Instrumentation: Instrumentation has the objective of providing means to perform and monitor the experiment in a way that does not affect its control and finally presents the same results regardless of how it is instrumented. The result of an experiment may be considered invalid if a new experiment is performed in the same application context that presents a divergence of results.

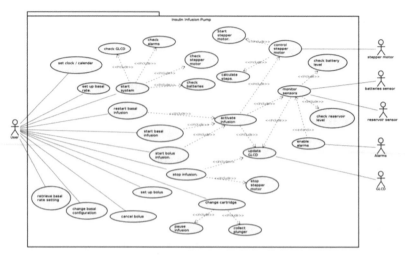

Figure 5.3   Insulin infusion pump requirements expressed in the use case diagram.

There are three types of instruments for an experiment: objects, guidelines, and measuring instruments. Objects are defined during the design of the experiment and can be, for example, requirements specification codes or documents. This experiment addresses the requirements specification document for an insulin infusion pump.

Guidelines for guiding experiment participants may include process descriptions and checklists, as well as training in the methods to be used by participants [19]. The measurements in this experiment were performed manually by collecting data that will compose a LibreOffice Calc spreadsheet.

The main guidelines used are the articles selected through a systematic literature review performed by the author in his master's dissertation. The authors show how each approach is applied and serves as objects of comparison between the selected approaches. However, in the context of the insulin pump, the requirements are modeled, and test cases and test behavior trees are generated for the analysis of the hypotheses described here.

C. *Operation*
In the operation phase of the experiment, the treatments were applied to the subjects. There are three steps to the operation phase: preparation, execution, and validation.

In the preparation phase of the experiment, the 28 requirements modeled here extracted from the Martins et al. [12] work infusion pump documentation, as shown in Figure 5.3, were raised.

This requirement model covers the three functionality modules of the insulin infusion pump system: user interface, infusion control, and sensor monitoring.

The user interface module comprises the functions related to the GLCD screen, control buttons, alarms, and system clock; Infusion control encompasses stepper motor functions; Sensor Monitoring is responsible for managing the system's insulin cartridge and battery check functions.

As per the specification, the requirements are presented in a use case diagram, and it was decided to organize better and identify them, to list them all in alphabetical order with an enumerated prefix keeping the same description given in the document. Table 5.2 lists the requirements.

The experiment was carried out in the context of the software of an insulin infusion pump by applying two model-based software requirements testing approaches. The approaches were selected based on an RSL [14] and applied to the system in question.

The modeling took place with the random selection of the requirements. Initially, the activity diagrams were created following the generation of the models; the descriptions of the test cases were made. The second modeling started with the behavior tree models and then the test behavior trees.

In the absence of computer-aided software engineering (CASE) tools available in the literature to model behavioral trees in a computerized manner, the whole modeling process was manually performed in both applied approaches.

After the modeling process in both approaches and the computed metrics, activity diagrams were transported to the free CASE Astah UML[1] and the behavior trees for the LibreOffice Impress[2] . This process was performed only to present well-designed images for a better understanding of the experiment and not for analysis purposes. All processes were performed by the master dissertation author.

In the validation step, the results of the experiment were statistically tested. The student's t-test application examined the rejection or non-rejection

---

[1]CASE tool for student modeling. Download at:http://astah.net/download

[2]A free program for editing and slideshows contains free design tools that enable the modeling. Download at: https://libreoffice.org/

Table 5.2   Requirements

ID	Requirement
R01	Drive Stepper Motor
R02	Change Baseline Programming
R03	Enable Alarms
R04	Enable Infusion
R05	Update GLCD
R06	Calculate Steps
R07	Cancel Bolus
R08	Configure Bolus
R09	Set Clock
R10	Set Basal Rate
R11	Stepper Motor Control
R12	Boot System
R13	Start Basal Infusion
R14	Start Bolus Infusion
R15	Monitor Sensors
R16	Stop Infusion
R17	Stop Stepping Motor
R18	Pause Infusion
R19	Collapse Plunger
R20	Retrieve Basal Rate Setting
R21	Restart Basal Infusion
R22	Change Cartridge
R23	Check Alarm
R24	Check Battery
R25	Check GLCD
R26	Check Stepper Motor
R27	Check Battery Level
R28	Check Reservoir Level

of the proposed hypotheses. This analysis is described in the results and analysis chapter of the experiment.

5.4 Results and Discussion

The results of the experiment were extracted based on the data collection worksheet and according to the execution processes informed in the previous section.

Conducting the experiment showed that in both approaches, 26 of the 28 requirements were modeled – these requirements were extracted from the insulin infusion pump requirements specification document.

Initially, it was noted that requirement R17 (Stop stepper motor) has redundancy compared to requirement R16 (Stop infusion). The specification document shows these two requirements in a similar way, which made the redundancy of R17 noticeable in one step of R16. For this reason, requirements R16 and R17 have been unified and named R16.

Then the requirements R24 (Check Battery) and R27 (Check Battery Level) were also unified for similar behavior, as both verify the existence of the battery and the contained power level to inform the user. Requirements R24 and R27 have been joined and named R27.

To meet certain functionality, some requirements need to be met by other software requirements. For example: to start an insulin infusion, we must also search for the patient profile that will receive the insulin.

Requirements are modeling through activity diagrams presented 11 requirements with this characteristic. They were called dependents in the sense of relationship with each other in meeting the specified functionality.

Table 5.3 describes the relationship between the source requirements and the requirements that contain a dependency on them. They are described as:

Req - Requirements;

Dep - Dependency;

Desc - Description .

Requirement behavior trees differently present the relationship between dependent requirements. First, the partial behavior tree is generated, it represents a requirement individually, and all behavior is described as part of it without reference to other requirements.

Then the partial trees are unified into a behavior tree that will show the whole system operation. This complete behavior tree shows at various points where one requirement needs to be met by another.

The modeling of insulin infusion pump software requirements was measured according to the metrics established in the experiment protocol. Data were recorded, and the effort in time in minutes used in each approach was verified.

The time effort applied during modeling activity diagrams was 144 minutes, for the generation of behavior trees were added 80 minutes. Figure 5.4 presents a graph with the time cost information for each modeled requirement.

Requirements R10, R12, R22, R23, and R25 are the requirements that required the most time effort in modeling activity diagrams. Due to the

Table 5.3   Description of dependency between requirements.

Req	Dep	Desc
R03	R05, R15	When executing a system alarm, it is necessary to update the GLCD display and keep sensor monitoring active.
R04	R11, R15, R05	The infusion activation process comprises a different set of functions and components.
R11	R06, R01	Stepper motor control requires calculating steps according to infusion rate and starting the motor.
R12	R25, R23, R26, R24, R05, R10, R04	When starting the system, the software checks the operation of its components.
R13	R04	Requirement R13 describes the time verification mode for infusion to start at the scheduled time.
R14	R08, R04	R14 requests and verifies the infusion data; if necessary, R08 allows the data to be saved, and finally, the infusion is started by R14.
R15	R28, R27	Sensor monitoring is by simultaneous verification through requirements R28 and R27.
R16	R05	R16 receives the command from the user and prompts the stepping motor to stop, so the information is displayed in the GLCD.
R21	R04	When paused, the infusion can be restarted through requirement R04.

Continued

Table 5.3 Continued

Req	Dep	Desc
R22	R18, R19	During cartridge change, the infusion in progress will be paused, and the plunger retracted first to position the motor.
R27	R05	At every battery level check, the battery icon is updated on GLCD.

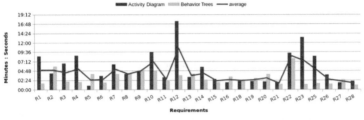

Figure 5.4 Graphical representation of time effort spent on requirements modeling.

complexity of these requirements, the total time devoted to them was 42.36% of the total time applied to the model.

For behavior tree modeling, the percentage of time spent on requirements R10, R12, R22, R23, and R25 is equivalent to 25% of the total time spent modeling. Behavior trees compared to activity diagrams presented the requisite models in a succinct and less detailed manner.

Looking at the time effort spent on requirements modeling is possible to see that the activity diagram with the least time spent was requirement R05 with 01 minute. It describes updating the icons on the system GLCD screen. In behavioral trees, the shortest time spent was 01 minutes at R28, which describes checking the insulin reservoir level.

While these requirements pertain to distinct use cases, time spent modeling shows that simple functionality tends to be a low cost of time. As an example of this statement, we have the requirements R28 and R26, and they got 1 minute time in the behavior tree and 2 minutes in the activity diagrams, respectively.

On the other hand, requirement R12 describing system startup has put more effort into the activity diagram with 18 minutes in modeling. In behavior trees, the most time spent was assigned to R22 with the description of the 8 minutes insulin cartridge change.

For the longest time spent, it can be seen that there was no proximity in the effort employed to the models. The activity diagram marked 14 minutes more than the behavior tree of that same requirement. In this context of insulin infusion pump software, the requirements with more details required more time for modeling activity diagrams.

Requirements that do not perform many data validations on the system tend to have fewer execution flows. In both the activity diagram and behavior tree, the amount of alternative flows that make up a requirement implies a more considerable modeling effort. The engineer modeled each situation presented by the requirement, thus increasing the time spent in the modeling process.

Similarly, requirements R02, R05, R16, and R27 belong to functions that only receive and send information. Sophisticated features such as system initialization (R12) perform a series of data send and receipts as well as validations and component checks, generating more time for modeling.

Activity diagrams detail software requirements based on the UML language, in which activity diagrams represent an external view of the system user. The tasks that the user/actor will perform are presented in activities and form primary and alternative flows.

The detailed requirements modeled through the BML language behavior trees show the software requirements in an external view of the complete system. The tasks that will be performed by each requirement are presented through states that the system can assume so that all system components can assume execution states in a behavior tree.

Among states and activities, it is noted that requirement R12 presented fewer behavior tree states than activities in the activity diagram. Thus, the time spent in their behavior tree was three times shorter than the time spent on the activity diagram.

Similar to the time used in requirement R12, requirements R3, R4, R10, and R23 were above the average time effort line for activity diagram modeling. Thus, it was possible to note that the abstraction of the actor actions who operates the software requires a greater understanding of the system operation. This situation is the opposite of the trees of behavior.

Test case generation for activity diagrams was performed manually. It occurred by writing the specification of the test cases, observing all the flows that make up the diagram, and pointing out the inputs and outputs of the

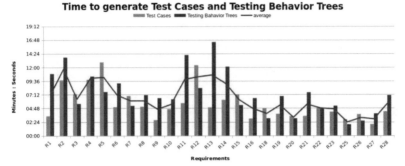

Figure 5.5   Graphical representation of time effort spent on test design

system. Each test case presented in the model was described step by step, as well as the pre and post conditions of the test case.

Test behavior tree generation was based on behavior tree models. Generated in manual mode, the test behavior trees have an assertion that must be satisfied by the test. With an assertion described, the test model points out the possible ways of executing the behavior tree to the expected result. The root node of the tree represents the test precondition, and the leaf nodes the postconditions according to each branch of the tree.

The calculated time was 145 minutes for test case generation and 193 minutes for tree generation and test behavior. An overview of the time spent creating use cases is shown in Figure 5.5.

Figure 5.5 shows 15 models corresponding to the requirements R1, R2, R6, R8, R9, R10, R11, R13, R14, R16, R19, R21, R23, R27, and R28, where test behavior trees presented higher cost in time of formation. This more expensive generation was due to the number of nodes that the tree and testing behavior can assume. That is, the more nodes to be modeled, the longer the time spent modeling.

During the generation of test behavior trees, according to the precondition (root node) defined, the child nodes that will generate n execution possibilities subtrees are added. Each subtree has a different path with different states, so behavior trees that have fewer tree paths generate less time effort on creation.

The specification of test cases obtained the most time effort on nine models, which correspond to the requirements: R03, R05, R07, R12, R15, R18, R20, R25, and R26. These test cases generated high effort due to the number of cases to be described. With natural language, the input values, situation description, and expected behavior of the test are written. However, the more cases to describe, the higher the cost in time. Coincidentally, it occurred that two requirement models (R04 and R22) obtained equal time for both approaches. These requirements models generated an effort of 10 and 05 minutes, respectively.

Noteworthy are the test behavior trees of the requirements models R1, R11, R13, R14, and R21, as they measure twice or more than twice as long as the test cases. These test behavior trees show more than one test branch for the assertion imposed by the requirement model.

In test cases, these requirements had the inputs and outputs described with the expected behaviors. As you can see in Figure 5.5, this did not cost much time as it was done in natural language; the test cases do not follow a flow to be specified.

In behavior trees, other branches are branched through an assertion to test the behavior of the functionality in question. The branches of a tree contain the data to be tested; this process required the engineer more modeling time because the higher the number of items to be tested, the more branches are created, increasing the effort of the engineer.

Unlike this scenario, use cases were not as costly, as seen in Figure 5.5. In a few functionalities, the test cases obtained high values for the test behavior trees. This statement is considered to the fact that the use case specification uses natural language.

Above all, generating test cases and test behavior trees is a task that required a deeper understanding of the functionality of the insulin infusion pump system. Figure 5.5 shows the average time considering the application of both approaches to each requirement model. The image shows that the generation of test cases through activity diagrams proved to be advantageous at creation time. However, the case written in natural language will not necessarily be better compared to behavior trees. At this point, what is under consideration is the time to generate the test case through the requirement models.

A. *Descriptive Statistical Analysis*

It is interesting to observe in Figure 5.4 the average time cost values between the approaches. In these values, a central time trend between both annotated values can be identified. For each modeled requirement, is the set of time values of approach (A), the time employed in each model (a), and the number of requirements (n), the formula that calculates the arithmetic mean is defined:

$$A = \frac{1}{n} \sum_{i=1}^{n} a_i = \frac{a_1 + a_2 + \ldots + a_n}{n}$$

By knowing the arithmetic mean time values of each requirement, one can observe according to the complexity of the functionality as each approach treats the model. The behavior tree approach presented below-average values for low complexity requirements, such as requirement R1, that start the engine sending the number of steps to perform. It is essential to consider, besides the arithmetic mean, a measure that indicates a margin of error for the processes performed. The standard deviation shows how time measurements in the model generation, test cases, and test behavior tree can vary. Thus, the standard deviation (SD) is determined by the square root of the time variance. The formula was used:

$$SD = \sqrt{\frac{\sum_{i=1}^{n} (x_i - M_a)^2}{n}}$$

As:

n - the size of the requirement set;

x_i - the time spent on each model, cases, and test trees;

M_a - the arithmetic mean of the dataset.

Through time, the time effort is subtracted from the mean arithmetic value and is squared, so divided by the amount of modeled requirements. The variance value of the set was obtained. The variance value shows how much the data set deviates from the mean. The calculation of the standard deviation showed, in minutes, an upper margin and a lower margin for time deviation calculated on each model, case, and test tree.

The effort measurement data in time are presented in Table 5.4. The values correspond to minutes. Caption:

BT - behavior tree;

TBT - test behavior tree.

TC - test cases;

AD - activity diagram;

AM - arithmetic mean.

Table 5.5 shows the following values: Total Time (TT), Minimum (MIN), Maximum (MAX), Arithmetic Mean (AM), Median (MD), Variance (VA), Coefficient of Variation (CV), and Standard Deviation (SD) of each process performed, the time is expressed in minutes.

Table 5.4   Effort in time employed in applying the approaches (in minutes).

Req	AD	BT	AM	Req	TC	TBT	AM
R01	9	2	5.5	R01	3	11	7
R02	4	6	5	R02	10	14	12
R03	7	2	4.5	R03	7	6	6.5
R04	9	2	5.5	R04	10	10	10
R05	1	4	2.5	R05	13	8	10.5
R06	4	2	3	R06	5	9	7
R07	7	4	5.5	R07	7	5	6
R08	4	4	4	R08	5	7	6
R09	5	5	5	R09	3	7	5
R10	10	5	7.5	R10	5	6	5.5
R11	3	2	2.5	R11	6	14	10
R12	18	4	11	R12	12	8	10
R13	3	4	3.5	R13	5	17	11
R14	6	3	4.5	R14	6	12	9
R15	3	2	2.5	R15	7	5	6
R16	2	3	2.5	R16	3	7	5
R18	2	2	2	R18	5	3	4
R19	2	3	2.5	R19	4	7	5.5
R20	2	4	3	R20	4	3	3.5
R21	2	1	1.5	R21	3	8	5.5
R22	10	8	9	R22	5	5	5
R23	14	1	7.5	R23	4	5	4.5
R25	9	2	5.5	R25	3	2	2.5
R26	4	1	2.5	R26	4	3	3.5
R27	2	3	2.5	R27	2	4	3
R28	2	1	1.5	R28	4	7	5.5
Total	**144**	**80**	-	-	**145**	**193**	-

Table 5.5   Approaches application time metrics (in minutes).

Process	TT	MIN	MAX	AM	MD	CV	VA	SD
DA	144	1	18	4	5.5	1.2	17.1	+-4.1
AC	80	1	8	3	3	0.9	2.8	+-1.6
CT	145	2	13	5	5.5	1.2	7.7	+-2.7
ACT	193	2	17	7	7.4	0.8	13.4	+-3.6

The smaller the variance value, the closer to the mean the values are. In the modeling process, 57.69% (15) of the activity diagrams that exceeded the average modeling time in behavior trees were 26.92% (07). In test cases, 34.61% (09) were above average in test behavior trees 57.69% (15).

However, it was observed that the requirements modeling process was less costly for behavior trees than for activity diagram modeling. The process of creating test cases proved to be less costly for test behavior trees.

To test the null hypotheses described in the protocol, we used the student t-test. The student t-test applies statistical concepts to reject or not a null hypothesis; it follows a t distribution under the null hypothesis [4, 11, 17]. According to the null hypotheses are $\bar{X} = \mu_0$ and consequently the alternative hypotheses $\bar{X}$, μ_0. The possibility of $\bar{X} < \mu_0$ and $\bar{X} > \mu_0$ [13, 19] is also being evaluated.

According to the characteristics described above, these represent a two-tailed test, in which a t-test will be applied to two population means, with two independent samples, with unknown population standard deviations. However, analyzing the values of Table 5.4, considering N the number of subjects in the table, we have N 1 + N 2 − 2 = 50; that is, the test will be performed. With degrees of freedom = 50 and with a significance level of 95%. This data provides a critical value for $t\,c = 1.67591$.

To determine the value of variance (S), we applied:

$$S^2 = \frac{\Sigma X^2}{N} - \bar{X}^2$$

AS:

S^2 - It is the square of the sum of time;

N - is the number of subjects;

X^2 - is the square of the arithmetic mean of the dataset.

The combined variance of the two groups was then calculated - Activity Diagrams (S_1) X Behavior Trees (S_2) and Test Case Specification (S_1) X Trees of test behavior (S_2) respectively in each calculation, with the formula:

$$S_{\bar{x}_1 - \bar{x}_2} = \sqrt{\left(\frac{N_1 S_1 + N_2 S_2}{N_1 + N_2 - 2}\right)\left(\frac{N_1 + N_2}{N_1 x N_2}\right)}$$

Therefore, the value of the t-test can be calculated as:

$$t = \frac{\bar{X}_1 - \bar{X}_2}{S_{\bar{x}_1 - \bar{x}_2}}$$

To test at H_0, a unit of test t is defined, and a critical area C is provided as well; this is a part of the area over which t varies. This means that the significance test can be formulated as [19]:

- If $t \in C$, reject H_0;

- If $t \notin C$, not reject H_0.

Therefore, applying this procedure to null hypotheses 1 and 2, it was decided that:

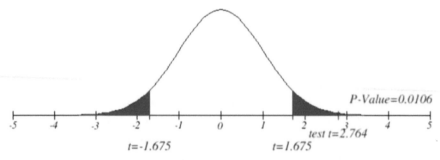

Figure 5.6   Function probability density student t of H_{01}

The t-test for H_{01} where $|t| = 2.7640 > t_c = 1.67591$, concludes that the null hypothesis is rejected. Looking at the P-value approach: the p-value is $p = 0.00106$, so $p = 0.00106 < 0.05$, thus reaffirming that the null hypothesis was rejected, so there is sufficient evidence to assert that population means are different. And there is significance. Figure 5.6 shows these data graphically.

In figure 5.6, the white area in the center of the graph represents the hypothesis acceptance zone. The distance between the value of t and the beginning of the hypothesis rejection point is marked by blue, while the red indicator shows the area in which the hypothesis was rejected. Figure 5.7 shows the graphical data for the second null hypothesis.

When applying student's t-test to H_{02} where $|t| = 2.0769 > t_c = 1.67591$, concludes that the null hypothesis is rejected. Looking at the P-value approach: the p-value is $p = 0.00482$, so $p = 0.00482 < 0.05$, thus reaffirming that the null hypothesis has been rejected, so there is sufficient evidence to assert that the population averages are different. And there is significance.

Hypotheses 3 and 4 are discussed in the following subsection, as well as other non-statistical aspects.

B. *Hypothesis Discussion*

In the protocol of the experiment, three hypotheses were defined to be tested with the collected data. These hypotheses may be considered accepted when the test result agrees with it and rejected when the test results do not agree with the hypothesis.

The null hypotheses H_{01} and H_{02} state the equality of generation time of requirements models and test cases. In these hypotheses, a test case is

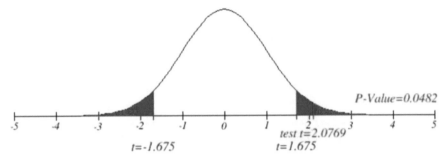

Figure 5.7 Function probability density student t of H_{02}.

considered as a set of conditions used to test the functionality of a software. In the context of the hypothesis, test cases, as well as test behavior trees and requirement models, describe condition sets and models for performing software requirements testing.

Modeling processes, as well as test case generation and test behavior trees, were performed manually, and the effort in time was noted in minutes. As seen in Table 5.4, the data calculated from the time effort employed in these processes.

The sum of the time of each approach presented in the Table 5.4 footer shows that the generation of test cases through the activity diagrams promoted in less time cost the generation of test behavior trees. The specification followed the diagram activity flows. To write a test case, all flows were described, and at each decision point, there was a new written test case.

The time difference from test case generation over test behavior trees was 48s minutes less for test cases. With this difference, it has been found that it is faster to write what is modeled, rather than modeling a test behavior tree from a behavior tree, i.e., H_{02}: Cost_Time (TC) < Cost_Time (TBT).

The process of generating test case specifications also served to point out defects in activity diagrams. Some decision points not covered in the diagrams could be noticed and corrected during the specification of the test cases, so it was assumed that the requirement models are correct.

However, according to the data presented, it was found that the creation of test cases through activity diagrams obtained the most extended time recorded in 9 requirement models. Test behavior trees with 15 of the most cost-effective requirement models; only two models are presented at the same time in both approaches.

According to the null hypothesis 3 raised in the experiment protocol, H03: Amount(BTB) = Amount(TC). The insulin infusion pump requirements document presented 28 software requirements, from which 26 models were generated. From this process, 26 test cases and test behavior trees were generated by both approaches.

For the context of this experiment, these numbers show the acceptance of null hypothesis 3, as they match the results expected by the hypothesis. Thus, it is possible to state that one approach generated the same number of cases and tests as the other.

The completeness of the models allowed us to explore all primary and alternative flows in activity diagrams, as well as all behavior tree nodes. Each decision point was clearly described within the requirement models, which provided the functionality of the requirement in the model.

The hypothesis H_{04} states: Amount(RAD) = Amount(RBT). It was found that the same amount of requirements were modeled for both approaches. The requirements described in the use case diagram provided by the requirements specification document allowed the two approaches to be joined together. The redundant requirements were uniquely interpreted by each approach.

Acceptance of the H_{04} hypothesis shows that these approaches, while different, are capable of identifying requirement redundancies, as well as improving traceability through model components and requirement completeness.

Models described through activity diagrams have fewer components that aid in requirement traceability. Behavior trees have symbols that more clearly indicate the origin and destination of a requirement.

In a complete behavior tree, nodes can be colored to indicate important actions; in partial behavior trees, there is no need to color because the information it contains is described in the complete behavior tree. Inactivity diagrams, tasks can be separated by streaks that demonstrate who performs each step. Therefore, hypothesis H_{04} was accepted as expected. However, according to the characteristics of the software engineering experiment, this experiment should have the same results written here when replicated.

C. *Advantages and disadvantages*
Each approach has its particularities, but it is still necessary to observe how each one stood out during the execution of the experiment. We evaluated both the requirements models and the test cases/test behavior trees.

The requirement models expressed in activity diagrams presented the highest sum of effort in time applied to the model. However, these diagrams show the functionality of the requirements clearly and objectively; the decision points contain descriptions of the actions that will be taken and the next flow path. This feature makes it easy for system stakeholders to understand the model.

In behavior trees, the model understanding has a different view of an activity diagram. It shows the response for each state while executing a requirement. Behavior trees colorfully show the type of requirement that will be fulfilled, and traceability operators more clearly show the requirement and its source.

Although behavior trees have less effort in certain use cases, it is not unanimous that it will be less costly for requirements than an activity diagram. As already seen, several requirements have gotten less effort into activity diagrams rather than behavior trees.

The specification of the test cases presented less effort in time applied in the experiment. The definition of natural language proved to be less expensive. With this approach, it was noted that the specification of written test cases better abstract the requirement. Testing behavior trees through tree branches show the ways to perform a test, in which the reading of the model can become tiresome because it does not contain a described but modeled explanation.

It should also be noted that behavior trees and test behavior trees have more components for requirements traceability. These components allow the relationship between requirements to be identified, such as => indicates that the requirement passes execution control to the requirement pointed to by the symbol.

There are also markings such as the + symbol, which indicates that the marked node is a requirement improvement during the modeling process. Just as the − symbol indicates that the node has been inserted into the tree and is not in the documented requirement. This aspect makes it possible to state that behavior tree modeling has better traceability than activity diagrams.

However, the experiment showed that it is simpler to view an activity diagram rather than a complete behavior tree. However, an activity diagram shows the user's view when using the system, while a behavior tree presents a more technical view, where nodes represent the states that the system and stakeholders can assume within the system.

5.5 Conclusions

The execution of the experiment showed that it took more time to model activity diagrams than to model behavior trees. Similarly, test behavior trees were more expensive than test cases.

It was also observed that due to the author having familiarity with activity diagrams, the result was not influenced; that is, it was expected that the activity diagram modeling would have a lower cost in time, but the results showed that this did not occur. The hypotheses raised in the experimental protocol were discussed, and the advantages and disadvantages found during the execution of the experiment were presented.

The application of the selected approaches aimed to show two different views on how modeling, requirements testing, and the generation of software test cases for the insulin infusion pump have occurred. Besides generating the obtained results, the adopted processes helped to identify problems in the requirements that could be corrected even in the modeling process. However, the approach that uses UML diagrams stood out in terms of cost-time in specifying test cases since behavior trees have a lower time-cost in modeling. All test cases (specification and models) represented the test possibilities described in each requirement model.

A. *Threats to Validity*
The experiment carried out has some threats to validity, among which we can highlight:

- The approaches found in the systematic literature review provided several methods for the experiment; according to the protocol raised, only two approaches were applied in the experiment. Applying other approaches can generate new results that add value to the results obtained.
- The experiment was carried out in the laboratory, limiting the execution of the requirements tests, as could be done in a case study. However, there was no test of the requirements models; the results of the experiment can be changed with the execution of the test.
- The context of the assessment did not include a specialist in the area of requirements testing by means of behavior trees. The participant in the assessment was a non-specialist researcher (master student), who, although trained in analysis and development, has no experience in requirements testing, which may result in different perceptions of the evaluation of the results obtained.

5.6 Future Work

There are works that can be done in the future after this experiment, they are:

- Run the test of requirements models generated in this work, based on the test cases and test behavior trees;
- Experiment replication in other test approaches based on requirements models;
- Conduct an experiment within the same context as the experiment with more work teams and compare results;
- Perform a case study for best results by conducting hands-on tests outside the lab.
- Perform tests to verify data distribution.

References

[1] Berenbach, Brian; Paulish, Daniel; Kazmeier, Juergen; Rudorfer, Arnold. Software &Amp; Systems Requirements Engineering: In Practice. New York, NY, USA: McGraw-Hill, Inc., 2009. ISBN 0071605479, 9780071605472.

[2] Cartaxo, Emanuela G.; Neto, Francisco G O.; Machado, Patricia D L. Test case generation by means of UML sequence diagrams and labeled transition systems. In: Conference Proceedings – IEEE International Conference on Systems, Man and Cybernetics (2007), pp. 1292–1297. ISBN 1424409918.

[3] Delamaro, Marcio; Jino, Mario; Maldonado, José. Introdução ao teste de software. Bd. 1. Elsevier Brasil, 2017. ISBN 9788535226348.

[4] Dodge, Yadolah. The concise encyclopedia of statistics. Springer Science & Business Media, 2008.

[5] Guedes, Gilleanes T. A. UML 2 - Uma abordagem prática. Novatec Editora, 2011. ISBN 9788575222812.

[6] Gutiérrez, J.J.; Escalona, M.J.; Mejı́t'as, M. A Model-Driven approach for functional test case generation. In: Journal of Systems and Software, vol. 109 (2015), Nov, pp. 214–228. URL http://linkinghub.elsevier.co m/retrieve/pii/S0164121215001703. ISBN 0164-1212

[7] Hammer, Theodore; Rosenberg, Linda; Huffman, Lenore; Hyatt, Lawrence. Measuring requirements testing. In: Proceedings - International Conference on Software Engineering (1997), pp. 372–379. – URL http://www.scopus.com/inward/record.url?eid=2-s2.0-030615

511{&}partnerID=40{&}md5=da901941f8e96376011c6117517186d f. –ISBN 0-89791-914-9.

[8] Hasling, Bill; Goetz, Helmut; Beetz, Klaus. Model-Based Testing of System Requirements using UML Use Case Models. In: International Conference on Software Testing, Verification, and Validation, IEEE, Apr 2008, pp. 367–376. – URL http://ieeexplore.ieee.org/document/45395 64/. ISBN 978-0-7695-3127-4

[9] Lamsweerde, Axel van. Requirements Engineering: From System Goals to UML Models to Software Specifications. Wiley Publishing, 2009. – ISBN 0470012706, 9780470012703

[10] Legeard, B.; Harman, M.; Muccini, H.; Schulte, W.; Xie, T. Model-based Testing: Next Generation Functional Software Testing. In: Practical Software Testing: Tool Automation and Human Factors (2010), pp. 1–13. – URL http://drops.dagstuhl.de/opus/volltexte/2 010/2620/. ISSN 1862-4405

[11] Mankiewicz, Richard. The story of mathematics. Cassell, 2000.

[12] Martins, Luiz Eduardo G.; Cunha, Tatiana; Oliveira, Tiago D.; Casarini, Dulce E.; Colucci, Juliana A. Documento de Especificação de Requisitos Bomba de Infusão de Insulina (protótipo), Unifesp. 2014.

[13] Oliveira, Andréia F. Testes estatzt'sticos para comparação de médias. In: Revista Eletrônica Nutritime vol. 5, n. 6, 2008. pp. 777–788.

[14] dos Santos, J., Martins, L.E.G., de Santiago Júnior, V.A. et al. Software requirements testing approaches: a systematic literature review. Requirements Eng 25, 317–337 (2020). https://doi.org/10.1 007/s00766-019-00325-w.

[15] Sarwar, Tabinda; Habib, Wajiha; Arif, Fahim. Requirements based testing of software. In: 2013 Second International Conference on Informatics & Applications (ICIA), IEEE, Sep. 2013, pp. 347–352. – URL http://ieeexplore.ieee.org/document/6650281/. – ISBN 978-1-4673-5256-7

[16] Sommerville, Ian. Engenharia de software. Tradução Ivan Bosnic e Kalinka G. de O. Gonçalves; revisão técnica Kechi Hirama. São Paulo: Pearson Prentice Hall, 2011. ISBN 978-85-7936-108-1

[17] Student. The probable error of a mean. In: Biometrika (1908), pp. 1–25.

[18] Wendland, Marc-florian; Schieferdecker, Ina; Vouffo-Feudjio, Alain. Requirements-Driven Testing with Behavior Trees. In: 2011 IEEE Fourth International Conference on Software Testing, Verification and

Validation Workshops, IEEE, March 2011, pp. 501–510. – URL http: //ieeexplore.ieee.org/document/5954455/. ISBN 978-1-4577-0019-4

[19] Wohlin, Claes; Runeson, Per; Höst, Martin; Regnell, Bjorn; Wesslén, Anders. Experimentation in Software Engineering: An Introduction. Springer Science, 2000. ISBN 9781461370918

6

Requirements Engineering in Aircraft Systems, Hardware, Software, and Database Development

J. C. Marques, S. H. M. Yelisetty and L. M. S. Barros

Instituto Tecnológico de Aeronáutica
Praça Marechal Eduardo Gomes, 50, São José dos Campos, São Paulo, Brazil
E-mail: johnny@ita.br; sara.mhy@gmail.com; lilian@ita.br

Abstract

Aviation is a safety-critical and regulated environment. Many requirements at different levels appear during the development of the aircraft and its systems. The objective of this work presents the types of requirements and standards involved in this domain for the system, software, hardware, and database. The primary aviation standards addressed in this work are SAE ARP 4754A, RTCA DO-178C, RTCA DO-254, RTCA DO-297, and RTCA DO-200B. Additionally, it is presented the main characteristics of how requirements engineering is conducted at different levels, including how certification requirements are organized.

Keywords: Aviation, System, Software, Hardware, Database, Requirements, Certification.

6.1 Introduction

Typically, safety-critical products are developed in regulated environments by standards. Examples are found in aviation, automotive, medical, railway, space, and nuclear. Examples of safety-critical systems are medical devices, nuclear power equipment, satellites, aircraft, and vehicles.

Standards are published by committees, technical entities, or regulatory agencies. They influence the development of safety-critical systems utilizing guidelines for systems, hardware, software processes, and products [1], considering the risks. Typically, each domain has its system and software standards, such as RTCA DO-178C [2] for aviation and IEC 62304 [3] for medical devices.

According to Thayer and Dorfman [4], Requirements Engineering is concerned with establishing and documenting software requirements. Additionally, there are some significant issues of concern in the requirements field: a) The inability of some engineers to write correct software requirements; b) The desire of managers to minimize requirements activity because they believe that the significant effort is programming; c) The lack of cooperation among team members, and; d) The lack of requirements' development methods.

According to Rierson [5], there are five reasons for the importance of reasonable requirements:

- Reason 1 - Requirements are the foundation for software development;
- Reason 2 - Good requirements save time and money;
- Reason 3 - Good requirements are essential for safety;
- Reason 4 - Good requirements are necessary to meet customer needs; and
- Reason 5 - Good requirements are necessary for testing.

Except for Reason 3, the other reasons impact any software development, not only safety-critical software.

According to the Federal Aviation Administration report: "Requirements Engineering Management Findings" [6], investigators focusing on safety-critical systems have found that requirements errors are most likely to affect the safety of an embedded system than errors introduced by design or implementation.

This work aims to present which types of requirements and standards are involved in aviation for systems, hardware, software, and database development to be used and certified as part of an aircraft project.

6.2 Aviation Standards

In aviation systems development, there are five standards typically required to be used as part of the aircraft certification process: SAE ARP-4754A [7], RTCA DO-178C [2], RTCA DO-254 [8], RTCA DO-297 [9], and RTCA DO-200B [10]. Figure 6.1 presents the relationship among these standards.

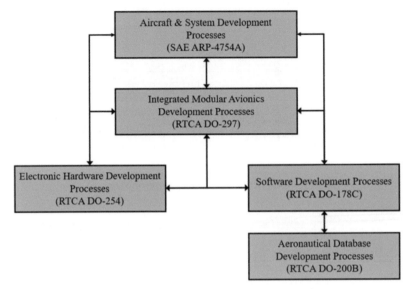

Figure 6.1   Relationship among aviation standards.

6.2.1 SAE ARP 4754A

According to Xiaoxun et al. [11], the SAE ARP-4754A [7] has been prepared primarily for electronic systems, which may be complex and readily adaptable to high levels of integration by their nature. However, the guidance is also applicable to engine systems and related equipment. It provides updated and expanded guidelines for the processes used to develop civil aircraft and systems that implement aircraft functions.

The SAE ARP-4754A has 47 objectives. These objectives are organized into 8 processes: Planning; Aircraft and System Development Process and Requirements Capture; Safety Assessment; Requirements Validation; Implementation Verification; Configuration Management; Assurance; and Certification Authority Coordination. These processes are organized in a process model, as presented in Figure 6.2.

6.2.2 RTCA DO-297

According to RTCA DO-297 [9], Integrated Modular Avionics (IMA) is a shared set of flexible, reusable, and interoperable hardware and software resources. When integrated, form a platform that provides services designed and verified to a defined group of safety and performance requirements to

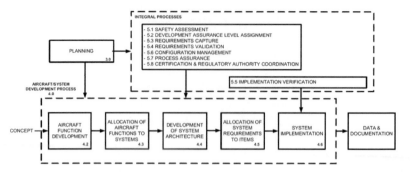

Figure 6.2   SAE ARP 4754A processes organization.

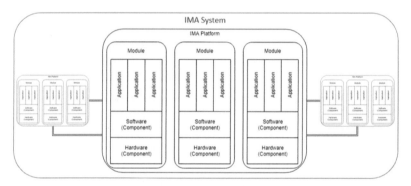

Figure 6.3   Levels of IMA approval.

host applications performing aircraft functions. An IMA system architecture is composed of one or more platforms and includes interfaces to other aircraft systems and users.

The IMA typical approach consists of the following levels of acceptance: Component; Module; Application; Platform; IMA System; and IMA Aircraft. These levels of acceptance are organized in Figure 6.3.

A component is a self-contained hardware part, software part, database, or configuration-controlled combination. A Component does not provide an aircraft function by itself. A Module may be software, hardware, or a combination of hardware and software, which provides resources to the hosted applications. Modules may be distributed across the aircraft or may be co-located.

An application is a collection of software and hardware modules with a defined set of interfaces that function when integrated with a platform. At Component, Module, and Application levels, the developments of such parts should follow RTCA DO-178C and RTCA DO-254. A Platform is a group of modules that establishes a computing environment, support services, and platform-related capabilities, such as health monitoring and fault management. An IMA System-level consists of a platform and a defined set of hosted applications. The IMA Aircraft-level should demonstrate that each aircraft function and hosted application functions as intended, supports the aircraft safety objectives, and complies with the applicable regulations. However, during the installation activities, the interactions between hosted applications relative to the provided aircraft functions should be verified and validated during aircraft ground and flight testing.

6.2.3 RTCA DO-178C

The early 1990s were characterized by a rapid increase in the extensive software usage in aircraft, engines, and airborne equipment [5]. This trend has resulted in the industry's need to create its guide and regulatory material to drive software development. The RTCA DO-178C [2] satisfied these needs. It guides the aeronautical community on software development processes [5].

The RTCA DO-178C has five software levels broken down into objectives to be satisfied. The satisfaction of applicable objectives enables the software approval as part of the aircraft certification process. Among the five existing Software levels (A, B, C, D, and E), level A is the most rigorous and requires compliance with all objectives. Level E refers to software products that malfunction without losing safety margins [12].

Each system failure is classified into five categories: Catastrophic, Hazardous, Major, Minor, and No Safety Impact. According to the most critical system failure condition, software products that malfunction cause or contribute to a system failure occurrence have an attributed software level. As presented in Table 6.1, the classification of the failure condition is associated with defined software levels. For each Software level, a set of objectives are required for compliance demonstration.

The 71 DO-178C objectives are presented in 10 tables, published in Annex A of the standard. The tables identify software process objectives with the following characteristics:

- Planning (Table A-1);
- Development (Table A-2);
- Verification of the high and low-level requirements and software architecture (Tables A-3, A-4, and A-5);
- Verification of source and executable codes (Tables A-5 and A-6);

Table 6.1   RTCA DO-178C software levels and number of required objectives [13].

Failure Condition Category	Software Level	Number of Required Objectives
Catastrophic	A	71
Hazardous	B	69
Major	C	62
Minor	D	26
No Safety Effect	E	0

- Testing and analysis (Table A-7);
- Configuration control (Table A-8);
- Quality assurance (Table A-9); and
- Certification (Table A-10).

The DO-178C provides recommendations for embedded software development and certification in civil aircraft, describing "what" shall be done and not "how" shall be done. In its structure, the various processes of the software life cycle are described through activities and objectives that shall be met. A table summarizes its objectives for each process, sections of the standard applied, and artifacts elaborated as evidence. Figure 6.4 shows the relationship among the various tables of DO-178C, in other words, the relationship among its processes.

The planning process defines and coordinates activities of development processes and integral processes. The software development process produces the product through the following steps: high-level requirements specification, architecture definition (design), specification of requirements more detailed (Software Low-Level Requirements) based on the Software High-Level Requirements (SW-HLR), development of source code based on Software Low-Level Requirements (SW-LLR), software tests and integration. The integral processes control the development processes; ensuring verification of generated data, ensuring the quality of processes by managing the configuration of evidence generated and during certification, ensuring compliance with the objectives, and generating visibility for certificate authorities. Additionally, the DO-178C has considerations such as reuse, alternative methods, and others. Table 6.2 shows the number of objectives per process.

As part of the effort of the DO-178C release, other supplementary standards were developed, including unique recommendations regarding tools qualification (RTCA DO-330 [14]), model-based development (RTCA

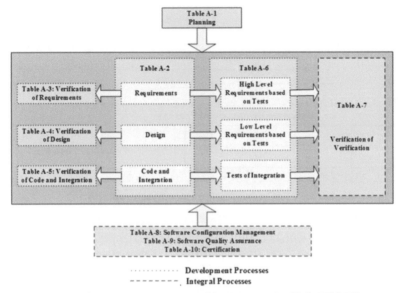

.... Development Processes

-------- Integral Processes

Figure 6.4 Relationship among tables (processes) of DO-178C [5].

Table 6.2 DO-178C tables and number of objectives [13].

Table	Number of Objectives	Processes
A-1	7	Planning
A-2	7	Development
A-3	7	Verification of Requirements
A-4	13	Verification of Design
A-5	9	Verification of Source Code and Integration
A-6	5	Tests of Integration
A-7	9	Verification of Results of Verification
A-8	6	Management of Configuration
A-9	5	Quality Assurance
A-10	3	Certification

DO-331 [15]), object-oriented technology (RTCA DO-332 [16]), and formal methods (RTCA DO-333 [17]).

6.2.4 RTCA DO-254

According to Kounish et al. [18], the RTCA DO-254 [8] is the standard used for the airborne and safety-critical application providing proper guidance to assure the design of Airborne Electronic Hardware. The RTCA DO-178C

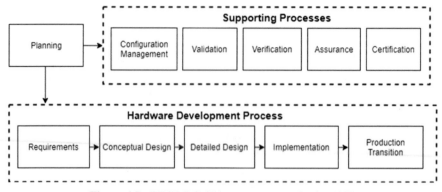

Figure 6.5   RTCA DO-254 processes organization [19].

has many similarities with RTCA DO-178C. For example, the RTCA DO-254 has five Design Assurance Levels (DALs) broken down into objectives to be satisfied.

The RTCA DO-254 objectives are presented in Table A-1, published in Appendix A of the standard. The RTCA DO-254 has 6 processes: Planning; Requirements; Conceptual Design; Detailed Design; Implementation; Validation; Verification; Configuration Management; Assurance; and Certification. These processes are organized and presented in Figure 6.5 [19].

6.2.5 RTCA DO-200B

Aeronautical Data is used for navigation, flight planning, flight simulators, terrain awareness, and other purposes. This standard provides guidance to assess compliance and determination of the levels of process assurance and supports the development of Aeronautical Databases.

According to the RTCA DO-200B [10], an Aeronautical Data Chain is a conceptual representation of the path that a set or element of aeronautical data takes from its origination up to its end-use. The RTCA DO-200B establishes three Data Process Assurance Level (DPAL) as the level of rigor representing the amount of verification and validation tasks performed during data processing to assure data quality.

For applications integrated into aircraft, the required DPAL is identified based upon the overall system architecture by allocating risk determined by using a preliminary system safety assessment, as specified in Table 6.3. Typically, the Aeronautical Data Chain involves many organizations. Data

Table 6.3 Failure condition categories and associated DPAL [10].

Failure Condition Category	DPAL
Catastrophic	1
Hazardous	
Major	2
Minor	
No Safety Effect	3

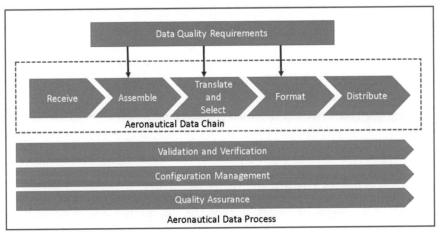

Figure 6.6 Aeronautical data process [20].

Providers are the organizations responsible for the data generated by them. Data Processors are the organizations accountable for using data from Data Providers and developing their data.

All data used to generate Aeronautical Database must meet Data Quality Requirements (DQR) specified by the Data Processor.

Figure 6.6 presents a typical Aeronautical Data Process with the following 5 phases: a) Receive; b) Assemble; c) Translate and Select; d) Format and; e) Distribute.

The Receive phase involves identifying and selecting sources of data that can support the Aeronautical Database Development.

The Assemble phase involves the collection and collation of data from one or more suppliers. It should result in a database that will meet the DQR of the next activity in the chain.

The Translate/Select phase involves the changes in how information is expressed. Checks ensure if the integrity of the original data after translation. The appropriate data is also selected if needed.

The Format phase involves converting the selected data subset into an acceptable format to the following functional link in the chain. It may take the shape of a published exchange standard format for the transmission of data, a proprietary format for loading in a target applications, or another agreed format. Checks are made to ensure that the data elements are compatible with the selected format. The source of every error is identified so that appropriate corrective actions can be taken.

The Distribute phase completes the processing data model and forms part of the transmission link in an aeronautical data chain. It involves the delivery of the formatted data set to users. Checks are carried out to ensure that the distributed data meets or maintains the DQR and no media errors exist. If errors or omissions are identified, they are reported to the appropriate participant in the data chain. Procedures are followed to ensure that the deficiencies are corrected and recorded for potential notification to the data end-users.

6.3 Requirements Engineering in Aviation

An aircraft is a system of systems. Typically, an aircraft has systems to perform its primary functions. Examples of Systems are Airconditioning, Auto flight, Communication, Electrical power, Landing gear, Flight controls, Navigation, Fuel, Oxygen, Hydraulic, Indicating/Recording, and Engine.

A System may contain subsystems. A subsystem may be composed of IMA System or standalone equipment. Each item is a software or a hardware part that performs some of the system functions. Figure 6.7 presents the organization among the type of parts of an aircraft.

There are many types of requirements associated with aviation system development. This work will briefly describe the following types of requirements:

- Certification Requirements;
- Aircraft and System Requirements;
- IMA System Requirements;
- Software Requirements; and
- Onboard Database Requirements.

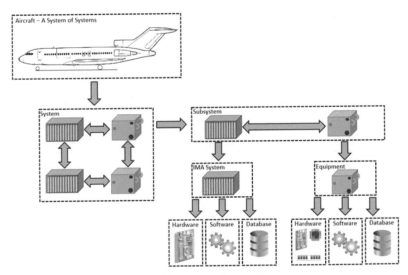

Figure 6.7 Types of aircraft parts.

6.3.1 Certification Requirements

Many safety-critical industries have some type of approval or certification process. For the aviation industry, certification is a demanding process that requires continuous attention. Failures to address the certification requirements in the project result in no certification or delays, making the project too expensive [5].

The term certification applies to aircraft, engines, and propellers, and some certification authorities consider auxiliary power units. Certification authorities as FAA (Federal Aviation Administration) and EASA (European Aviation Safety Agency) are responsible for aircraft certification in civil aviation. Software is considered part of the systems or equipment installed in the aircraft. In other words, the software is not considered an isolated product. Systems and equipment, including embedded software, shall be approved to be accepted as part of the aircraft certification.

As presented in Table 6.4, the aircraft certification requirements have categories (denominated as PART). Each part contains a set of certification requirements for different types of projects.

Although the standards presented in Section 6.2 are not predefined certification requirements, they are acceptable as a means of compliance by the FAA and other relevant certification authorities. Table 6.5 shows the Advisory Circulars (AC) that recognize each standard described herein.

Table 6.4   Categories (PART) and scope of certification.

PART	Scope
23	Small Aircraft
25	Transport Aircraft
27	Small Helicopters
29	Transport Helicopters
31	Balloons

Table 6.5   Advisory circulars and standards recognized.

AC	Standard
20-115D [21]	RTCA DO-178C
20-152 [22]	RTCA DO-254
20-153B [23]	RTCA DO-200B
20-170 [24]	RTCA DO-297
20-174 [25]	SAE ARP 4754A

6.3.2 Aircraft and System Requirements

According to Figure 6.8, Requirements Capture and Validation are essential processes associated with safety, certification, and implementation.

Activities associated with the Requirements Capture include:

- Aircraft-level functions, functional requirement, functional interfaces, and assumptions are defined;
- System requirements and system interfaces are defined;
- System architecture is defined;
- Derived System Requirements are identified and justified; and
- System requirements are allocated to the items.

Requirements Validation is the process of ensuring that the specified requirements are sufficiently correct and complete so that the product will meet the needs of customers, users, suppliers, maintainers, and certification authorities, as well as aircraft, system, subsystem, and item developers.

Activities associated with the Requirements Validation should ensure the requirements are:

- Correctly stated, feasible, identifiable, and unique;
- Not in conflict;
- Feasible;
- Complete; and
- Verifiable.

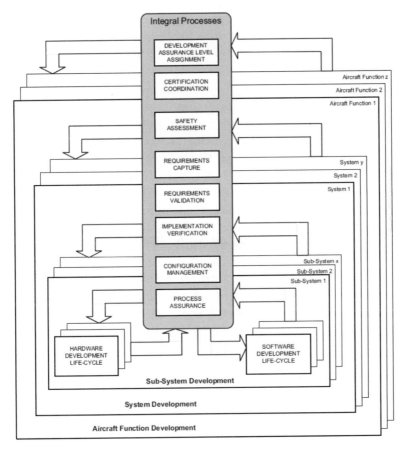

Figure 6.8   Aircraft function implementation process [7].

When using IMA, the system development process should address the following objectives:

- Identify aircraft functions, including functional, performance, safety, availability, and integrity requirements; and
- Determine any aircraft function requirements driven by the capabilities of the IMA platform modules, for example:
  - Availability requirements beyond those available from a single IMA module or platform which could drive the application to be hosted on multiple modules and platforms;
  - Applications using multiple modules should determine application redundancy management requirements; and

– Integrity requirements beyond those available from an IMA module or platform could drive the hosting of multiple instances of the application and data to achieve the necessary integrity.

6.4 Software Requirements

The RTCA DO-178C has a significant number of objectives associated with Software Requirements development, using as input System Requirements that Software will implement. There are two levels of Software Requirements on RTCA DO-178C. Software High-Level Requirements (SW-HLR) generally represent "what" should be designed. SW-HLRs include functional, performance, interface, and safety-related requirements. The Software Low-Level Requirements (SW-LLR) represent the how-to, providing details on implementing Software in code [26]. SW-LLRs include the features required for source code development, such as data coupling and control features.

The rationale for two levels of Software Requirements is the need to provide traceability and refinement from System Requirements to the level of implementation in source code. RTCA DO-178C requires the definition of a Software Requirements Standards (SRSt), which shall define the methods, notations, rules, and tools to develop the SW-HLRs, which shall be adherent to SRSt.

Activities associated with the development of Software High-Level Requirements and Software Low-Level Requirements include:

- Each allocated System Requirement for Software must be specified in Software High-Level Requirements;
- Each Software High-Level Requirement must be refined into Software Low-Level Requirements;
- Each Software Requirement (HLR and LLR) must adhere to the Software Requirements Standards (SRSt) and be verifiable and consistent;
- Each Software Requirement (HLR and LLR) shall be established in quantitative terms with tolerances, where applicable; and
- Each derived Software Requirement (HLR and LLR) must have a justifiable reason for its existence.

The Software High-Level Requirements and Software Low-Level Requirements review should ensure that:

- The Software High-Level Requirements are traceable and compliant with System Requirements;

- The Software Low-Level Requirements are traceable and compliant with Software High-Level Requirements;
- The Software Low-Level Requirements are accurate and consistent;
- The Software Low-Level Requirements are compatible with the computer environment;
- The Software Low-Level Requirements are verifiable, possible to provide any evidence of satisfaction; and
- The Software Low-Level Requirements are compliant with Software Requirements Standards (SRSt).

The Software architecture is developed from the Software High-Level Requirement. Additionally, the manufacturer shall establish and document the architecture, including the interfaces between internal and external components.

6.4.1 Model-Based Software Requirements

The Model-Based Development (MBD) can be used to specify requirements at many levels: Aircraft, System, Software High-Level Requirements (SW-HLR), or even Software Low-Level Requirements (SW-LLR).

Typically, the development of software using modeling is defined as Model-Based Development (MBD). As presented before, the RTCA DO-178C defines two levels of requirements (HLR) and (LLR). The RTCA DO-331 [15] introduced some definitions:

- Specification Model – A model representing Software High-Level Requirements that provides an abstract representation of functional, performance, interface, or safety characteristics of software components. A Specification Model does not define software design details such as internal data structures, internal data flow, or internal control flow; and
- Design Model – A model that defines any software design such as Software Low-Level Requirements, software architecture, algorithms, component internal data structures, data flow, and control flow. A model used to generate Source Code is a Design Model.

There are many possibilities to use models to define Software High-Level Requirements, Software Architecture, and Low-Level Requirements; some options are:

- Specification Model as Software High-Level Requirements (Figure 6.9 (a));

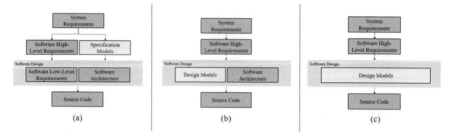

Figure 6.9   Some possibilities of use of model-based development.

- Design Model as Software Low-Level Requirements (Figure 6.9 (b)); and
- Design Model as Software Low-Leve Requirements and Architecture (Figure 6.9 (c)).

6.4.2 Software Requirements Using Object-Oriented Technology

According to RTCA DO-332 [16], Object-oriented technology (OOT) has been widely adopted in non-critical software development projects. The use of this technology for critical software applications in avionics has increased. Still, several issues need to be considered to ensure the safety and integrity goals are met. These issues are both directly related to language features and to complications encountered with meeting well-established safety objectives.

OOT's key features and related techniques are inheritance, parametric polymorphism, overloading, type conversion, exception management, dynamic memory management, virtualization, traceability, structural coverage, component-based development, and resource analysis. Although Requirements Engineering is involved in all essential features, RTCA DO-332 specifies details associated with how software requirements are used in the design using classes, subclasses, and methods. Figure 6.10 can be summarized as:

- Class hierarchy should be developed based on Software High-Level Requirements;
- Class diagram is an artifact of the Software Architecture; and
- All functionalities are implemented in methods; therefore, Software Low-Level Requirements must be created to specify each method and attribute.

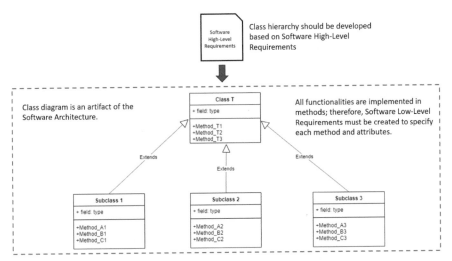

Figure 6.10 Use of OOT in requirements specification.

6.4.3 Software Requirements Using Formal Methods

According to RTCA DO-333 [17], formal methods are descriptive notations and analytical methods used to construct, develop, and reason mathematical models of system behavior. This standard also defines a formal analysis as using mathematical reasoning to guarantee that properties are always satisfied by a formal model. A formal model is a model described using a formal notation.

A property is a representation of a software requirement in a formal notation. Typically, a formal notation is precise, unambiguous, with syntax and semantics mathematically defined. Therefore, a formal method is an analysis carried out on a formal model.

Since RTCA DO-178B (1992), Formal Methods were considered as an alternative method that can be used to satisfy any objective. In RTCA DO-178C, after supplementing RTCA DO-333, Formal Methods are used as an acceptable means of compliance.

Different formal models may be used for different kinds of analysis throughout the development life cycle and the establishment of other properties. Examples of formal models include:

- Textual-Based Models (Figure 6.11 (a)); and
- Graphical Models (Figure 6.12 (b)).

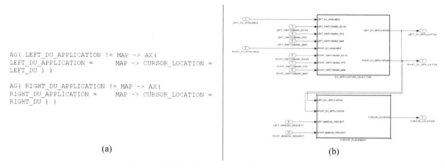

```
AG( LEFT_DU_APPLICATION != MAP -> AX(
LEFT_DU_APPLICATION =     MAP -> CURSOR_LOCATION =
LEFT_DU ) )

AG( RIGHT_DU_APPLICATION != MAP -> AX(
RIGHT_DU_APPLICATION =    MAP -> CURSOR_LOCATION =
RIGHT_DU ) )
```

(a)   (b)

Figure 6.11   Examples of formal models.

Activities associated with the usage of Formal Methods for requirements specification include reviews and analyses. The topics should include:

- Formal analysis cases: The objective is to provide evidence that the formal analysis covers the review and analyses of Software High-Level Requirements, Software Low-Level Requirements, and Software Architecture;
- Formal analysis procedures: The objective is to verify that the formal analysis cases were accurately developed into formal analysis procedures and expected results; and
- Formal analysis results: The objective is to ensure that the formal analysis results are correct and those discrepancies between actual and expected results are explained.

6.5 Hardware Requirements

According to RTCA DO-254 [8], the hardware requirements capture process identifies and records the hardware item requirements. It includes those derived requirements imposed by the proposed hardware item architecture, choice of technology, the primary and optional functionality, environmental and performance requirements, and the requirements imposed by the system safety assessment. This process may be iterative since additional requirements may become known during design.

The objectives for the requirements capture process are:

- Requirements are identified, defined, and documented. It includes allocated needs from the PSSA and derived requirements from the hardware safety assessment; and

- Derived requirements produced are fed back to the appropriate process.

The validation process is not intended to validate the requirements allocated from system requirements since validation of these requirements is assumed to occur as part of the system process, using SAE ARP 4754A [7]. The objectives of the validation process for derived hardware requirements are:

- Derived hardware requirements against which the hardware item is to be verified are correct and complete; and
- Derived requirements are evaluated for impact on safety.

6.5.1 Onboard Database Requirements

The database is a set of data that influences the software's behavior without modifying the executable code and that is managed as a separate item during the architecture definition of a system product [27]. There are two types of embedded databases in aviation: Parameter Data Items (PDI) and Aeronautical Databases [28].

6.5.2 Parameter Data Items

Parameter Data Items are usually approved under the type design of an aircraft. These databases are also part of the systems and may influence paths during code executions. They also can be used to activate or deactivate software components and functions by adapting software computations to the aircraft configuration or being used as computational data. The Parameter Data Items (PDI) are typically approved under the RTCA DO-178C [2].

The definition of PDI Requirements uses the values defined by the System Requirements to be included in the Parameter Data Item. Additionally, the attributes and structure of the Parameter Data Item must be refined from the Software Requirements and Design. The PDI Requirements are equivalent to Software High-Level Requirements (HLR) [20], and they should define how the PDI affects the function of the software.

According to RTCA DO-248C [29], the PDI structure, attributes, and values should be defined in requirements to allow requirements-based verification of the PDI File. Since High-Level Requirements are required for all software/assurance levels, the decision was made to define structure, attributes, and values in the high-level requirements.

The actual values in the PDI are considered high-level requirements that are allocated to the PDI only. As presented in Figure 6.12, these may trace directly to system-level requirements or maybe considered derived high-level requirements.

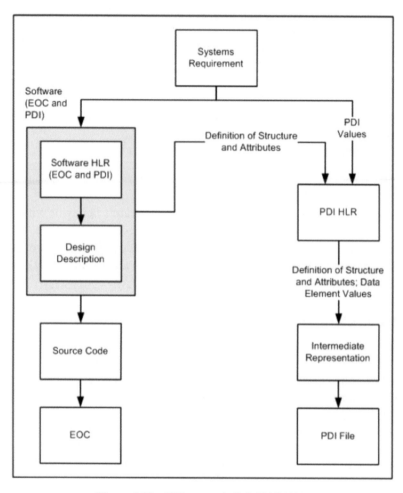

Figure 6.12   PDI process in DO-178C [29].

6.5.3 Aeronautical Databases

Aeronautical Databases are typically not approved under the type design of an aircraft. An airborne system uses these databases, and development processes generally are approved using the guidance of the RTCA DO-200B [10]. Aeronautical Databases are limited to navigation, terrain, obstacle, and airport map. All data used to generate Aeronautical Database must meet Data Quality Requirements (DQR) specified by the Data Processor. The Data Quality Requirements shall characterize the data by:

- Accuracy;
- Resolution;
- Confidence that the data has not been corrupted while stored, processed, or transmitted (assurance level);
- Ability to determine the origin of the data (traceability);
- Level of confidence that the data applies to the period of the intended use (timeliness); and
- Format.

6.6 Conclusion

The objective of this work was to present which types of requirements and standards are involved in aviation for systems, hardware, software, and databases development to be used and certified as part of an aircraft project.

Section 6.2, it was summarized the primary aviation standards associated with systems, software, hardware, and databases: SAE ARP 4754A [7], RTCA DO-178C [2], RTCA DO-254 [8], RTCA DO-297 [9], and RTCA DO-200B [10].

Section 6.3, it was presented the main characteristics of how requirements engineering is conducted at different levels. It was summarized: a) how certification requirements are organized; b) Aircraft and System Requirements; c) Software Requirements, including the use of Model-Based Development and Formal Methods; d) Hardware Requirements, and; e) Onboard Database Requirements.

References

[1] J. Munch, O. Armbrunt, M. Kowalczyk, M. Soto, 'Software Process Definition and Management', Springer-Verlag, Berlim, Germany, 2012.
[2] Radio Technical Commission for Aeronautics, 'DO-178C Software Considerations in Airborne Systems and Equipment Certification', Washington, USA, 2011.
[3] International Electrotechnical Commission, 'IEC 62304: 2015 Medical Device Software-Software Life-Cycle Processes', 2015.
[4] R. H. Thayer, M. Dorfman, 'Software Requirements Engineering', 2nd Edition, IEEE Computer Society, 2000.
[5] L. Rierson, 'Developing Safety-Critical Software: A Practical Guide for Aviation Software and DO-178C Compliance', CRC Press, 2013.

[6] D. L. Lempia, S. P. Miller, 'Report DOT/FAA/AR-08/34: Requirements Engineering Management Findings', Washington, USA, 2010.

[7] Society of Automotive Engineers, ARP-4754A Guidelines for Development of Civil Aircraft and Systems, 2011.

[8] Radio Technical Commission for Aeronautics, 'DO-254 Design Assurance Guidance for Airborne Electronic Hardware', Washington, USA, 2000.

[9] Radio Technical Commission for Aeronautics, 'DO-297 Integrated Modular Avionics (IMA) Development Guidance and Certification Considerations', Washington, USA, 2011.

[10] Radio Technical Commission for Aeronautics, 'DO-200B Standards for Processing Aeronautical Data', Washington, USA, 2015.

[11] L. Xiaoxun, Z. Yuanzhenb, F. Yichenb, S. Duoa, 'A Comparison of SAE ARP 4754A and ARP 4754', Procedia Engineering, Vol. 17, pp: 400–416, 2011.

[12] J. Marques, A. Cunha, 'A Reference Method for Airborne Software Requirements', Proceedings of the 32nd IEEE/AIAA Digital Avionics Systems Conference (DASC), Syracuse, USA, 2013.

[13] S. M. H. Yelisetty, J. Marques, P. M. Tasinaffo, 'A Set of Metrics to Assess and Monitor Compliance with RTCA DO-178C', Proceedings of the 34th IEEE/AIAA Digital Avionics Systems Conference (DASC), Prague, Czech Republic, 2015.

[14] Radio Technical Commission for Aeronautics, 'DO-330 Software Tool Qualification Considerations', Washington, USA, 2011.

[15] Radio Technical Commission for Aeronautics, 'DO-331 Model-Based Development and Verification Supplement to DO-178C and DO-278A', Washington, USA, 2011.

[16] Radio Technical Commission for Aeronautics, 'DO-332 Object-oriented and Related Technologies Supplement to DO-178C and DO-278A', Washington, USA, 2011.

[17] Radio Technical Commission for Aeronautics, 'DO-333 Formal Methods Supplement to DO-178C and DO-278A', Washington, USA, 2011.

[18] R. Koushik, M. Anushree, B. J. Sowmya, N. Geethanjali, Design of Spi Protocol with DO-254 Compliance for Low Power Applications. Proceedings of the 2017 International Conference on Recent Advances in Electronics and Communication Technology (ICRAECT), 2017.

[19] D. S. Loubach, J. C. Marques, A. M. da Cunha, 'Considerations on Domain-Specific Architectures Applicability in Future Avionics

Systems'. Proceedings of the 10th Aerospace Technology Congress (FT), Stockholm, Sweden, 2019.

[20] J. Marques, A. Cunha, 'Verification Scenarios of Onboard Databases under the RTCA DO-178C and the RTCA DO-200B', Proceedings of the 36th IEEE/AIAA Digital Avionics Systems Conference (DASC), St. Petersburg, USA, 2017.

[21] Federal Aviation Administration, 'Advisory Circular 20-115D Airborne Software Development Assurance Using EUROCAE ED-12() and RTCA DO-178()', Washington, USA, 2017.

[22] Federal Aviation Administration, 'Advisory Circular 20-152 Document RTCA/DO-254, Design Assurance Guidance for Airborne Electronic Hardware', Washington, USA, 2005.

[23] Federal Aviation Administration, 'Advisory Circular 20-153B Acceptance of Aeronautical Data Processes and Associated Databases', Washington, USA, 2016.

[24] Federal Aviation Administration, 'Advisory Circular 20-170 Integrated Modular Avionics Development, Verification, Integration and Approval using RTCA/DO-297 and Technical Standard Order C153', Washington, USA, 2013.

[25] Federal Aviation Administration, 'Advisory Circular 20-174 Development of Civil Aircraft and Systems', Washington, USA, 2011.

[26] J. C. Marques, S. M. H. Yelisetty, L. A. V. Dias, A. M da Cunha, 'Airborne Software Certification Accomplishment Using Model-Driven Design', International Journal of Advanced Computer Science, Vol. 3, No. 1, Pp. 18-25, 2013.

[27] M. Hernandes, 'Database Design for Mere Mortals: A Hands-On Guide to Relational Database Design', Addison-Wesley Professional, 2013.

[28] Federal Aviation Administration, 'Order 8110-49 Software Approval Guidelines Change 2', Washington, USA, 2017.

[29] Radio Technical Commission for Aeronautics, 'DO-248C Supporting Information for DO-178C and DO-278A', Washington, USA, 2015.

7

Generating Safety Requirements for Medical Equipment

A. Martinazzo, L.E.G. Martins and T.S. Cunha

Federal University of São Paulo (UNIFESP), Institute of Science and
Technology, Talim, 330 - Vila Nair, São José dos Campos, SP, Brazil
E-mail: martinazzoaldo@gmail.com; legmartins@unifesp.br;
ts.cunha@unifesp.br

Abstract

Patient and operator safety has outstanding relevance during the use of
medical equipment. To address safety aspects, certification of medical
equipment in many countries requires conducting the risk management
process defined in ISO 14971. This process includes identifying risks
associated with medical equipment use, evaluating the acceptability of these
risks, and implementation of risk control measures as required. Even certified
equipment may present safety issues in operation, including recalls. Lessons
learned from problems in operation are used to improve risk management in
new developments. In addition, better results are achieved if risk management
is conducted in parallel with equipment development. However, standard ISO
14971 does not provide integration between risk management and equipment
development activities.

Therefore, a framework was developed using requirements engineering to
integrate risk management and equipment development processes during the
development stage of the product life cycle. This integration is performed
in such a way that the risk management process influences equipment
development. From equipment architecture to detailed definitions, every
design decision considers safety requirements from the risk management
process. This approach precludes design changes of great impact driven

by safety reasons. The framework can also contribute to achieving better coverage of hazards identification and reduction of adverse events and recalls.

Keywords: Framework, medical equipment, ISO 14971, risk management, safety requirements.

7.1 Introduction

Safety of patient and operator has fundamental importance during the use of medical equipment. To protect patients and the health team, certification of medical devices requires risk management. ISO 14971 is the standard adopted to perform risk management for certification of medical products in many countries [1]. However, even certified equipment presents safety-related issues during operation, which requires a continuous effort to maintain an adequate level of protection.

Analysis of recalls and adverse event reports provide information about safety problems encountered during equipment operation. A study was conducted to determine the number of medical devices subject to recalls or warnings in the UK. during the period from January 2006 to December 2010. The recalls were issued by the device manufacturer and published by U.K. Medicine and Health Regulatory Authority (MHRA). Additionally, MHRA sent a medical device alert to end-users of equipment. The study found 2,124 recalls, with the number of recalls per year increasing from 2006 to 2010. From these recalls, 15% were related to high-risk devices, with a prevalence of cardiovascular systems and muscle-skeletal systems, 75% were medium-risk devices, and 12% low-risk devices. This study also identified 447 medical device alerts, 44% related to the risk of death or serious health problem [2].

Another study about medical device recalls was conducted in the United States from January 2005 to December 2009. A correlation between the severity of recalls and the FDA approval process to release the device to market was performed. The rigor of this process is defined by the regulatory classification of the device, which is based on the amount of regulatory control considered necessary to ensure safe operation. The study found 13 recalls with a risk of death or severe adverse health effect. Most of the 113 devices were originally released by FDA using the less stringent process [3, 4].

A study to identify software-related recalls from 2005 to 2011 was conducted by FDA. This study analyzed 5,792 medical device recalls and found that 1,112 (19.4%) of them were related to software. The expansion of software utilization with increasing complexity in medical equipment explains the growth of software-related recalls [5].

FDA evaluated adverse events and recalls of infusion pumps. Problems encountered include software errors, issues related to interface with the user, and components failures. Some of these problems resulted in excessive or insufficient medication infusion. FDA found that many of the problems encountered were caused by the development errors that could be avoided. Guidance material was issued by FDA in 2014, aiming to improve the quality of infusion pumps and preclude adverse events and recalls. According to this guidance, performing safety assurance cases simultaneously with product development enhances the quality and prevents expensive and time-consuming reworks for compliance with the safety requirements [6]. However, standard ISO 14971 does not address synchronization between the risk management process and the medical device development process [7]. Therefore, this chapter aims to present a framework using requirements engineering to integrate risk management and equipment development processes.

The remainder of the chapter is organized as follows: Section 7.2 describes related works; Section 7.3 presents an overview of standard ISO 14971 and describes the framework for risk management simultaneous with medical equipment development; Section 7.4 includes conclusions.

7.2 Related Works

Researchers of the FDA defined the model that represents the behavior of many insulin-infusion pumps. This model was used to prepare the hazard analysis for a generic insulin pump. The identification of hazards was performed with the participation of manufacturers, pump users, and clinicians. In addition, adverse event reports were used to elicit hazardous situations. The hazard analysis resulting from this work is organized in hierarchical levels of hazardous situations linked to each other. This organization permits to use the generic hazard analysis as a reference during the entire development of insulin-infusion pumps, starting with the high-level hazardous situations and deploying into deeper details as the design progresses [8].

The growing number of insulin pump functions implemented by software brings complexity and increases the possibility of design errors. The same group of FDA researchers used the generic hazard analysis as input to generate generic safety requirements for insulin pump software to address this risk. The set of requirements defined in this study is organized by functions implemented by software, and each requirement is traced to the hazard analysis. Prevention of software errors depends on the quality of requirements set and the rigor of the software development process [9, 10].

7.3 Framework for Integration of Risk Management Process with Medical Equipment Development

This section presents an overview of the risk management process according to ISO 14971 and then a detailed narrative of the framework which integrates risk management with medical equipment development.

7.3.1 Risk Management Process According to ISO 14971

An overview of the risk management process according to ISO 14971 is depicted in Figure 7.1. The process starts with identifying hazardous situations, defining their effects, and corresponding severity classification. Hazardous situations are defined by ISO 14971 as *"circumstances at which person, property or environment are exposed to danger"* [7].

The next step is to verify if the risk posed by each hazardous situation is acceptable according to the established risk acceptance criteria. Risk control measures need to be provided for hazardous situations in which risk is considered unacceptable. However, the risk control measures themselves introduce additional risks that need to be analyzed [7].

Safety-related information gathered during the production and utilization stages of life cycle is used to verify whether the conclusions of risk management conducted for certification remain valid in actual operation scenario. This verification includes: checking whether the information from operational scenarios includes hazardous situations not considered for certification and verifying whether analysis of collected information modifies the severity classification and acceptability of hazardous situations [7].

Information from the actual operation is also useful to improve the risk management process for new developments [7].

The framework described in this chapter covers the development stage of the medical equipment life cycle. Therefore, the activities related to production and utilization stages are not within the framework scope.

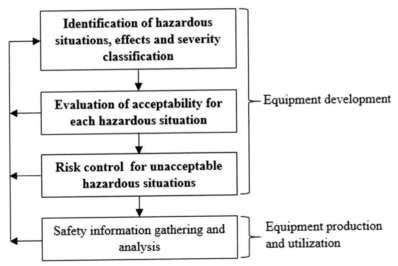

Figure 7.1   Overview of risk management process according to ISO 14971 (adapted from [7].)

7.3.2 Framework Description.

A framework integrating ISO 14971 risk management and equipment development processes during the development stage of the product lifecycle is shown in Figure 7.2. Definition of hazardous situations and corresponding evaluation of risk acceptability is accomplished in two main levels of detail, synchronized with the evolution of equipment development [7, 11–13].

The hazardous situations level 1, identified in an early step of equipment development, support the definition of equipment architectural design. Evaluation of these hazardous situations verifies whether the architecture is acceptable from the point of view of risk management and defines safety requirements for detailed design. Risk control measures are applied as required to architectural design aiming to reach acceptable risk [7, 11–13].

The hazardous situations level 2 are defined at a deeper level of detail, aiming to support the development of hardware and software components and verify that detailed design satisfies risk acceptance criteria [7, 11–13].

Hazardous situations can result from several causes that include user errors, software development errors, and electromagnetic disturbances [7]. Therefore, the framework provides integration with the standards that address these causes: standard IEC 62366 related to usability; standard IEC 62304 about software development process; and standard IEC 60601-1-2, which covers electromagnetic disturbances.

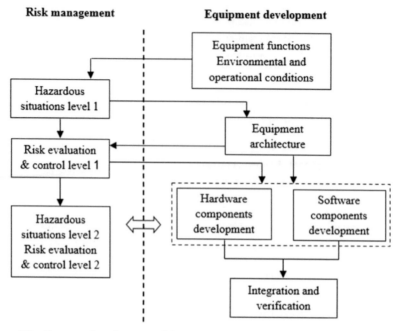

Figure 7.2　Framework to integrate risk management and equipment development (based on [11]).

The following subsections describe each box of the framework represented in Figure 7.2, emphasizing those related to the risk management process.

7.3.2.1 Equipment Functions

Equipment functions with associated requirements and the intended environmental and operational conditions are defined according to inputs from medical device stakeholders. These definitions are the starting point for the development of medical equipment integrated with risk management [11, 12].

7.3.2.2 Hazardous Situations Level 1

Activities related to hazardous situations level 1 are performed at an early phase of development, before the definition of architectural design, and are conducted using the information available at this moment. The objective is to identify hazardous situations at a high level of abstraction, define the effects on the health of the patient and medical crew, and classify these effects

Table 7.1 Severity classification of hazardous situations [10].

Severity classification	Effects on patient
Critical	*Death, injury with risk of death or permanent disability*
Moderate	*Reversible or moderate injury*
Negligible	*Mild injury or absence of injury*

according to their severity, aiming to establish safety requirements for the definition of equipment architecture [7, 11].

The inputs to perform this analysis are equipment functions with associated requirements; intended environmental and operational envelope; checklist for safety characteristics of medical devices (Annex C of ISO 14971); and reports of adverse events and recalls for the same type of equipment [7, 11]. The outputs are the list of hazardous situations level 1 and effects with severity classification for each one [7, 11]. These outputs are used to define the equipment from the safety point of view.

Identification of hazardous situations level 1 is independent of the implementation and is based on:

- Failures of equipment functions [14]:
  - Loss of function, which is easier to deal with.
  - Erroneous behaviors of the function, which are more difficult to identify completely.
- Concerns specific to medical devices such as toxicity, infection, allergic reaction, and compatibility with medical tissues [7].
- Reports of adverse events and recalls occurred during the use of the same type of equipment [7].

Hazardous situations level 1 identification, the definition of effects, and severity classification need to be performed and validated with the participation of a multidisciplinary team, including specialists from the medical area.

The rule to classify the severity of hazardous situations is defined by the equipment manufacturer [7]. The present framework adopted the severity classification from the standard for medical device software, presented in Table 7.1 [10]. A common definition for severity classification provides better integration between risk management and software development processes.

Hazardous situations result from an isolated event or a combination of events. These events can be caused by physical components (hardware failures), user errors, software-related errors, or environmental conditions, including electromagnetic disturbances [7].

Physical component failures can be random or systemic. Radom failure occurs when a physical component that was working properly no longer performs as intended. Systemic failure is caused by a development error that becomes evident in a specific operational situation or environmental condition. Systemic failures due to production errors are not in the scope of this study. Protection against physical component failures includes redundancy with adequate independence [7].

Electromagnetic fields can cause the malfunctioning of electronic components. Protection against electromagnetic interference is provided by shielding or filtering. Means to recover from a transient disturbance may also be adopted [15].

Utilization error is an erroneous action performed by the equipment user or lack of an action that results in a response not expected with a possible adverse effect on safety [16].

Software-related errors can be caused by flaws of input requirements or errors during software development itself.

The risk acceptance criteria for medical equipment are defined by the manufacturer and approved by the regulatory authority [7]. In this work, we will consider the criteria described as follows, based on ISO 14971 and other standards related to medical device safety:

- Hazardous situations with severity classification critical shall not result from a single failure [17].
- Whenever standards for medical equipment define acceptance criteria for specific hazardous situations, these can be adopted [7]. For example, IEC 60601-2-24 includes criteria for some hazardous situations of infusion pumps, and ISO 80601-1-12 provides guidance for certain failures of emergency care ventilators.
- Safety features to protect against specific hazardous situations shall be equivalent to the current practice on the same type of equipment [7].

The acceptance criterion for the risk arising from software development errors is defined by the standard IEC 62304—Medical device software, which establishes the level of rigor for software development according to software safety classification defined in Table 7.2.

Definition of equipment architecture from the safety perspective is based on hazardous situations level 1, their effects, severity classification, and corresponding risk acceptance criteria.

Table 7.2 Software safety classification [10].

Severity classification of software function failures	Effects of software error on patient	Software safety classification
Critical	*Death, injury with risk of death, or permanent disability*	C
Moderate	*Reversible or moderate injury*	B
Negligible	*Mild injury or absence of injury*	A

7.3.2.3 Equipment Architecture

The architecture is a high-level design of the medical equipment. The inputs do define architectural design include but are not limited to:

- Equipment functions and associated requirements.
- Description of environmental conditions for equipment use, which includes the electromagnetic environment.
- Characteristics of equipment users.
- Hazardous situations level 1 with associated effects and severity classification.
- Risk acceptance criteria for hazardous situations.

Definition of equipment architecture needs to be supported by risk management activities to verify that risk acceptance criteria for hazardous situations level 1 are satisfied. These activities are described in risk evaluation and control level 1. Safety features such as redundancy, partitioning, and monitoring are required for compliance with risk acceptance criteria for hazardous situations classified as critical [11, 14].

The output is the definition of architecture, which encompasses:

- The list of equipment components with associated functions and requirements.
- Description with a visual illustration of how the components are interconnected to perform equipment functions.
- Identification of equipment functions implemented in software.
- Preliminary design of user interface.
- Preliminary design of interface with patient.
- Environmental qualification requirements for the equipment.

7.3.2.4 Risk Evaluation and Control Level 1

Risk evaluation and control level 1 needs to be performed nearly simultaneously with the definition of architecture to validate architectural design from the safety point of view. Another objective is to define safety requirements for detailed design. The inputs to conduct this risk evaluation and control are [7, 11, 14]:

- Architecture candidates.
- Hazardous situations level 1 with associated effects and severity classification.
- Standards with safety recommendations for the equipment under development.
- Information from the literature about safety features adopted in devices of the same type.

The outputs of this activity are:

- Validation of architectural design from the safety aspect.
- Definition of safety requirements to support detailed design.

Validation of architecture from the safety perspective is performed by verifying that acceptance criteria for hazardous situations level 1 are satisfied, as described in the following paragraphs.

Compliance with risk acceptance criteria for hazardous situations with severity classified as critical requires redundancy. It is necessary to identify all failure combinations that cause hazardous situations level 1 with severity critical to demonstrate that redundancy exists. Fault Tree Analysis (FTA) visually presents the failure combinations. However, common cause failures may defeat redundancy. Therefore, an analysis aiming to ensure that redundancies are effective is conducted using information available from architectural design. This analysis starts with the identification of critical redundancies, as illustrated in Figure 7.3. The redundancies are represented by "and" gates in the fault tree [6, 11, 14].

Some safety standards for the product under development include risk acceptance criteria for specific hazardous situations. It is necessary to search for these criteria and to verify if equipment architecture complies with them.

Another acceptance criterion establishes that safety features to address specific hazardous situations shall be equivalent to the practice adopted on other equipment of the same type. Demonstration of compliance with this criterion demands a literature review to search for information about the safety features of existing devices.

In case of noncompliance with the criteria, it is necessary to apply risk control measures and revise the risk management [7].

Monitors and alarms are included in architectural design and can trigger an automatic action or an alarm for the equipment operator.

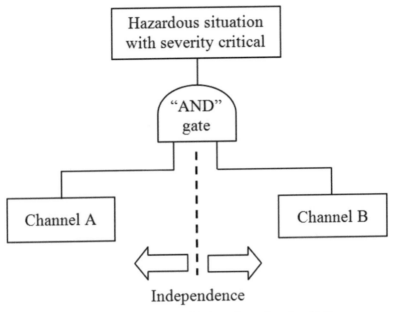

Figure 7.3   Redundancy and independence (based on [14]).

Failure analysis of monitors and alarms included in architectural design is carried out considering at least: loss of monitor/alarm activation when required; independence between monitor/alarm and monitored condition; and unintended activation of monitor/alarm.

A preliminary analysis of utilization errors is performed based on the user interface as defined in architectural design and high-level definition of user tasks. Errors during the accomplishment of each task are identified and described according to the level of detail of available information. The effects of user errors on patient health are defined and classified according to the severity. The contribution of these errors to each hazardous situation level 1 is mapped [16].

Software safety classification is conducted for functions implemented in software. This classification is used to define the level of rigor for software development process [10]. The steps to define software safety class according to Table 7.2 are:

- To identify all functions performed by software.
- To define the effect and severity of loss of functions performed by software and erroneous execution of the function.

- Consider the most severe effect of software function failure to define software safety classification.

A preliminary risk assessment related to electromagnetic disturbances is also performed and includes [15]:

- Identification of functions performed by electronic components susceptible to electromagnetic perturbations.
- Definition of effects and associated severity of failures of these functions, considering the loss of function and erroneous behavior of the function.

The set of safety requirements for detailed design encompasses:

- Requirements to ensure that redundancies have adequate independence.
- Identification of additional monitors and alarms required by safety reasons, with traceability to corresponding hazardous situations level 1.
- Software safety classification.

7.3.2.5 Development of Components

Safety requirements defined during risk evaluation level 1 and 2 are used for:

- Design of physical (hardware) components.
- Development of software components.
- Definition of monitors and alarms, which integrate hardware e and software.

7.3.2.6 Hazardous Situations Level 2 Evaluation and Risk Control

Identification and analysis of hazardous situations related to equipment components are conducted to verify adherence to risk acceptance criteria in deeper detail and define additional safety requirements to support components design.

The following activities are conducted to reach these objectives:

- Analysis of physical components failures.
- User errors analysis relative to safety.
- Safety analysis of monitors and alarms.
- Verification of adherence to the software development process.
- Electromagnetic disturbances.
- Detailed analysis of common cause failures.

Failures of physical components can be random or systemic. Protection against these failures is provided by redundancy and monitoring and a robust development process to preclude systemic failures. Compliance with risk acceptance criteria requires demonstration that severity classification of any single failure is not critical. Severity classification is defined by performing a bottom-up analysis that identifies all failure modes of components and describes the effects on the component itself, the equipment, and the patient. The results are cross-checked with top-down analyses (FTAs) conducted for hazardous situations level 1 with severity classification critical. Risk control measures such as redundancy and alarms need to be applied in case of noncompliance with the risk acceptance criterion. Systemic failures of physical components can be caused by requirements or design flaws. For example, an error in the definition of environmental requirements for the component can result in failure under specific environmental conditions [7, 11].

As the user interface design progresses, analytical and experimental methods are adopted to evaluate utilization error in further detail, with the participation of human factors specialists and equipment users. The analytical methodology includes the evaluation of tasks performed by the user during equipment operation. Experimental approaches are closer to actual operation and involve tests simulating device use [16].

The necessity of monitors and alarms for safety reasons can be identified in the following analyses: risk evaluation and control level 1; physical components failures; and user error analysis relative to safety. Safety analysis based on functions to be performed by monitors and alarms generates requirements for their design. This analysis considers the loss of monitor activation when required, undesired activation, and independence between the monitor and monitored condition. After monitoring/alarm design, a more detailed safety analysis is accomplished, including human factors considerations.

Software development needs to follow the process according to the safety classification defined in risk evaluation and control level 1. Verification of adherence to the process required for each software component is performed along with software development [10].

Risk management related to electromagnetic disturbances at this point of development encompasses [15]:

- Failures of components responsible for ensuring equipment immunity against electromagnetic interference.

• Definition of acceptable behavior of functions susceptible to electromagnetic interference during immunity tests.

Analysis of common cause failures is performed in further detail at this stage of development. Critical safety redundancies are challenged with deeper knowledge about potential threats that may degrade independence. The results of this analysis generate safety requirements to ensure acceptable risk for hazardous situations with critical severity [17].

7.4 Conclusion

The framework presented in this chapter provides a structure for the safety-driven design of medical equipment. Synchronizing risk management activities per ISO 14971 with the evolution of equipment development leads to design decisions supported by safety requirements since the initial development steps. The framework also provides an interface with these related processes: software development according to IEC 62304, usability defined by IEC 62366, and electromagnetic disturbances covered by IEC 60601-1-2.

Methodical application of the framework can contribute to achieving better coverage of hazardous situations identification, minimization of expensive design changes to correct non-conformities with safety regulations, improved patient and medical crew safety, and reduction of adverse events and recalls.

References

[1] A. Dolan, 'Risk management of Medical Devices: Ensuring Safety and Efficacy through ISO 14971,' Annual Quality Congress Proceeding, Milwaukee, 2004.

[2] C. Heneghan, M. Thompson, et al., 'Medical-device recalls in the U.K. and the device-regulation process: retrospective review of safety notices and alerts,' BMJ Open, 2011.

[3] D.M. Zuckerman, P. Brown, S. E. Nissen, 'Medical Device Recalls and the FDA Approval Process,' Arch Intern Med/Vol 171(No. 11) June 2011.

[4] Food and Drug Administration, 'Regulatory controls' available at https://www.fda.gov/medical-devices/overview-device-regulation/regulatory-controls accessed on April 15, 2020.

[5] L.K. Simone, 'Software-Related Recalls: An Analysis of Records,' Biomedical Instrumentation & Technology, November/December 2013.

[6] Food and Drug Administration, 'Infusion Pumps Total Product Life Cycle – Guidance for industry and FDA Staff,' December 2014.

[7] ISO 14971 'Application of risk management to medical devices,' 2019.

[8] Y. Zhang, P.L. Jones, R. Jetley, 'A Hazard Analysis for a Generic Insulin Infusion Pump,' Journal do Diabetes Science and Technology, vol 4, issue 2, March 2010.

[9] Y. Zhang, R. Jetley, P.L. Jones, A.Ray, 'Generic Safety Requirements for Developing Safe Insulin Pump Software,' Journal do Diabetes Science and Technology, vol 5, issue 6, November 2011.

[10] IEC 62304 'Medical device software,' 2015.

[11] SAE ARP4754A – 'Guidelines for Development of Civil Aircraft and Systems,' 2010.

[12] ISO/IEC/IEEE 15288 'Systems and software engineering – System life cycle processes,' 2008.

[13] ISO/IEC/IEE 24748-1 'Systems and software engineering – System life cycle processes – Part 1: Guidelines for life cycle management,' 2018.

[14] SAE ARP4761 'Guidelines and Methods for Conducting the Safety Assessment Process on Civil Airborne Systems and Equipment,' 1996.

[15] IEC 60601-1-2 – 'Medical electrical equipment – Part 1–2: General requirements for basic safety and essential performance – Collateral Standard: Electromagnetic disturbances – Requirements and tests,' 2014.

[16] Food and Drug Administration, 'Applying Human Factors and Usability Engineering to Medical Devices – Guidance for Industry and FDA Staff,' February 2016.

[17] IEC 60601-1 – 'Medical electrical equipment Part 1: General requirements for basic safety and essential performance.'

8

Meta-Requirements for Space Systems

C. H. N. Lahoz

Science and Space Technologies CTE, Institute Technological of Aeronautics
ITA, Sao Jose Campos, Brazil
E-mail: lahoz@ita.br

Abstract

This study, started by Marcos Romano, Carlos Lahoz, Edgard Yano in 2009, intends to summarize a set of meta requirements to use as reference for create any nonfunctional requirements computer system focused on space applications. These meta requirements presented in a hierarchical manner aims to help the requirements engineering team to better identify and discuss what is the minimum set of them that must be included in the system under analysis.

Keywords: Meta-Requirements, Space Systems.

8.1 Introduction

The space systems, which involve critical software, are increasingly complex due to the great number of requirements to be satisfied, which contributes to a higher probability of hazards and risks in a project. The software aims to make these systems easier to be implemented in a more integrated and functional way. However, because of such a huge number of components interacting dynamically, in many times with strong coupling, allows dysfunctional behaviors between the parts of the system.

Leveson [1] explains in her book that we are designing systems (complex ones) with potential interactions among the components that cannot be thoroughly planned, understood, anticipated, or guarded against. The

125

operation of some systems is so complex that it is a challenge to understand and be fully aware of their potential hazardous behavior.

The problem, according Leveson, is that this increasing interactive complexity makes it too difficult for us to manage. Software, for instance, due to its complexity and the coupling make it difficult for the designers to consider all its potential states, as well as, for the operators deal with all the situations (normal and abnormal) and disturbances in a safely and effectively manner. Besides she states that in the space project domain, the vast majority of software accidents were related to flawed requirements and misunderstandings about what the software should do [1, 2]. The generation of software requirements for critical systems is one of the major sources of errors in system development.

A manner of minimizing potential problems, related to malfunctions, accidents or threats to critical space systems, is to introduce a set of nonfunctional requirements to be considered in the requirement elicitation process at the system under development. The goal is promoting a deeper interaction between system and software engineers and improving the system quality and dependability requirements [3].

8.2 Requirements Engineering in Space Systems

In this section, issues related to problems in requirements engineering at space systems to provide the main motivation of this study are presented. After that, is shown a brief overview of the meta-requirements proposed in this study to use in the requirements elicitation process for space systems.

8.2.1 Requirements in Space Systems

The space systems automation increases considerably the complexity of the mission and the steps required to achieve it. Robots, rovers, and any kind of automation in space is becoming and has become an economic necessity when it comes to the new space and the progress into space. In the space systems projects, there is a gap between the requirements on software specified by systems engineers and the implementation of these requirements by software engineers [4]. The software engineers must perform the translation of the requirements into software code, hoping to accurately capture the systems engineer's understanding of the system behavior, which is not always explicitly specified. This gap opens up the possibility for the misinterpretation of the systems engineer's intent, potentially leading to software errors.

According to Hecht and Buettner [5], nearly half of all spacecraft anomalies observed in the period from 2 years (1998–2000), were related to software. The authors state that the study of requirements-originated software failures showed that roughly half resulted from poorly written, ambiguous, unclear, and incorrect requirements. The rest came from requirements that were completely omitted. The common reasons for the software failure in space system are related to inadequate practices of requirement specification and lack of knowledge of the system. In the investigation report of the Ariane 501 accident [6], from the European Space Agency (ESA) poor practices of requirement specification were identified. The investigation reports from Mars Climate Orbiter MCO [7] and Mars Polar Lander MPL [8], accidents, recommended a more adequate training of the development teams due to problems in the specification of the system's requirements. In the analysis of more than 18 ESA projects performed by Hjortnaes [9], he emphasizes the lack of maturity and stability of the baseline of software requirements when the development process begins. He also mentions that in many of the analysed reports, the development of the requirements was not conducted in an adequate way or there was not a correct understanding of them.

Leveson [2] described a set of spacecraft accidents related to software and the new paths for the hazards caused by the loss of information or by incorrect information leading to an unpredictable behavior of the space computer systems.

8.2.2 Meta-Requirements in Space Systems

As defined by Sommerville [21], functional requirements are statements of services the system should provide, how the system should react to particular inputs and how the system should behave in particular situations. The author also defines nonfunctional requirements as constraints on the services or functions offered by the system such as timing constraints, constraints on the development process, standards, and so on.

In this study, a set of nonfunctional requirements applied to space computer systems to be considered in the requirement process at the system under development was gathered.

Due to this set of requirements be characterized in a generic way, was chosen the meta-requirement label. Meta-requirements are not directly associated with a specific project or system requirement, only is associated with a set of criteria (or only recommendations) for meeting the implied requirements. For instance, the meta-requirement "robustness" does not specify features (physical or behavioral) related to a particular project. The meta-requirements are presented in abstracted way only to evaluate if they should be included or not in the system under analysis.

All of these meta-requirements, originally proposed by Romano et al. [3] area related to hardware, software, or both. Other contributions come from academic studies [11, 12], International standard organizations [13–16], as well as studies related to the dependability of some authors in the area [10, 17–21].

8.2.3 Requirement Engineering Process in Space Systems

The requirements engineering process is normally applied in the early phases of the software project life cycle but could be continues to be refined throughout the life cycle. This process is based on four iterative steps, resulting in a validated set of requirements that satisfies a set of stakeholder expectations. In general, the requirement engineering process covers the activities of elicitation, analysis, specification, and validation, as presented in Figure 8.1.

The requirements elicitation is the first step of requirements process and is used to building an understanding the system issues whose software is required to solve. In space systems, it is fundamentally a human activity and is where the stakeholder expectations are captured. It is usually accomplished through techniques like use-case scenarios, Design Reference Missions (DRMs), and Concept of Operations (ConOps). The second step is about the requirement analysis and consist to translate the expectations in high-level requirements used to drive an iterative loop where, for instance, the ConOps are derived into high level requirements. Third step, requirement specification, aims to achieve the consistency from the requirement analysis, creating a set of functional and nonfunctional requirements of the system. Finally, the fourth step is for the space project team to validate the

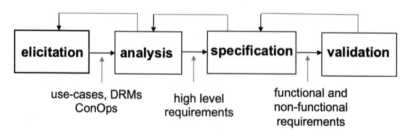

Figure 8.1   Requirement engineering process.

requirements against the stakeholder expectations. A simplified validation asks the questions: Does the requirements specifications identified cover how the whole system works? Is the nonfunctional requirements meet the system safe and reliable constraints? Are the requirements achievable within budget, performance, and schedule constraints?

8.3 Meta-requirements Selected to Space Systems

A summary of the meta requirements found and its brief definition is presented taking into consideration several authors' contributions in the requirement engineering field. Most of techniques to assure the meta-requirements are based on [3, 22–24].

The nonfunctional requirements proposed were classified in a product and process meta-requirements tree, taking into consideration some kind of dependency relationship between each other. The meta-requirements tree is strongly based on the quality factors proposed by [10] and divided into three branches: defensibility, soundness, and quality. Figure 8.2 shows the meta-requirements selected for space computer systems.

In the "defensibility" branch, space meta-requirements related to the way the system or its components can defend itself from accidents and attacks. In the "soundness" branch are space meta-requirements related to the way the system or its components is suitable for use. In the "quality" branch,

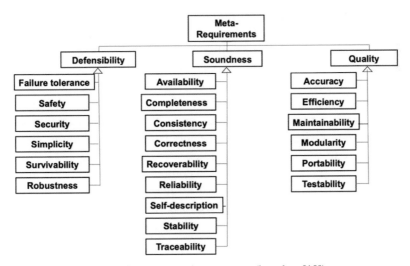

Figure 8.2   Meta-requirements tree (based on [10]).

other space meta-requirements related to quality assurance for product and for process to system or its components are defined.

Following, in alphabetic order, each meta-requirement selected for space computer systems is discussed.

8.3.1 Accuracy

Accuracy is the degree of the system presents correct or accurate answer resulting in mission accomplished. An inaccurate value resulting from the calculation of the logic of a spacecraft control may lead to insertion of errors, accumulated during its flight, leading it to follow an unexpected trajectory and the insertion of the satellite out of the desired orbit. An inaccurate value was one of the causes of the accident with Ariane 5 launcher in 1996 [1]. The precision of the navigation software in the flight control computer (On-Board Computer) depends on the precision of the Inertial Reference System measurements, but in the Ariane system's test facility this precision could not be achieved by the electronics creating the test signals. The precision of the simulation may be further reduced because the base period of the Inertial Reference System is 1 ms versus 6 ms in the simulation at the system test facility.

8.3.2 Availability

Availability is the degree of the space system is ready to carry out its task when need it to be. This attribute may be calculated as a function of Mean Time to Failure (MTTF) and Mean Time to Repair (MTTR). One example cited by [25] says that for a "service" type spacecraft such as the telephony/television communications satellite, down time, or "unavailability" constitutes loss of revenue, and hence the cost benefits of design improvements to increase reliability can be optimized against their impact on revenue return. As another example, the lack of navigation data during a certain period of time of the vehicle control cycle can destabilize it, in such a way to cause the loss of the mission. Therefore, subsystems or components of the vehicle as the on-board computer, the inertial system and the data bus should be available to perform their functions in the moment they are requested. A way to avoid failures and assure the availability of the computer system is to apply techniques such fault detection, fault recovery, and fault prevention. Examples of fault detection tactics are use ping/echo to request message between nodes, timestamps to detect incorrect sequences of events, condition monitoring, adoption of triple modular redundancy (TMR), and many others. Recovery faults included hot and cold spare, exception handling, revert to a previous stable state (rollback), retry the operation, reconfiguration, and so on. To fault prevention can be used code inspections, pair programming, reviews, and audits.

8.3.3 Completeness

Completeness is the degree to which the requirement satisfies the needs, goals, and/or mission of the space system. That means determining the stakeholders, the context where the system works and its capabilities to attend the mission goals. The accident report of the Mars Polar Lander, that was crashed during entry and landing stage in Mars in 2000, mentioned that in the requirements document, at the system level, did not specify the modes of failure related to possible transient effects to prematurely identify the touch of the ship on the ground. It is speculated that the designers of the software, or one of the auditors could have discovered the missing requirement if they were aware of its rationale [2]. This demonstrates that the nonconsideration of the completeness attribute in the requirements may lead to occurrence of a system failure. A way to improve the completeness in the requirement elicitation process for space systems is apply techniques and tools like requirements specification templates, requirements checklists, understand the system through use cases and others UML diagrams, create prototypes to validate the requirements or perform the Quality Function Deployment (QFD) method.

8.3.4 Consistency

Consistency is the capability of the space system specifications have no internal contradictions. During the American launcher Titan IV Centaur space accident investigation occurred in 1999, one of the causes found arose from the installation procedure of the inertial navigation system software, where the rolling rate -0.1992476 was placed instead of -1.992476. The fault could have been identified during the prelaunch, but the consequences were not properly understood, and the necessary corrections were not made because there was not a verification activity of critical data entry [1]. Keep a list of all measurement units (meters or inches, for instance) and scales of measurement (millimetres or meters, for instance) of the variables used in the system, the correct interval range of the variables, are some a way to help to assure the consistency in the requirements. Other satisfactory way to achieved consistency can be using a terminology pattern or a glossary understandable for all item or components of the system.

8.3.5 Correctness

Correctness is the capability of the space system or its components satisfy specific goals or to meet the overall business needs. Zowghi [26] states that in the formal point of view, correctness is usually meant to be the combination of consistency and completeness. In [27], it is reported that after having examined 387 software errors discovered during integration and system tests of the Voyager and Galileo spacecraft, it was concluded that most errors were due to discrepancies between the documented requirements specifications and the requirements necessary for the proper functioning of the system. Leveson [1] stated that in the Titan/Centaur accident, there was apparently no checking of the correctness of the software after the standard testing performed during development. For example, on the day of the launch, the attitude rates for the vehicle on the launch pad were not properly sensing the earth's rotation rate (the software was consistently reporting a zero-roll rate), but no one had the responsibility to specifically monitor that rate data or to perform a check to see if the software attitude filters were operating correctly. In fact, there were no formal processes to check the validity of the filter constants or to monitor attitude rates once the flight tape was actually loaded into the Inertial Navigation Unit at the launch site. Potential hardware failures are usually checked up to launch time, but it may have been assumed that testing removed all software errors, and no further checks were needed. A way to verify the correctness of the system can be by checking whether the outputs are consistent with the related requirements. It can be made by tests or even using other techniques such as inspections and walkthroughs.

8.3.6 Efficiency

Efficiency is the capability to boost performance or optimize resources allocation in the space system. Considering a real-time system, efficiency is a relevant attribute in the care of their temporal constraints, and is related to performance, as the checks from time response, CPU and memory usage. Examples of efficiency problems are the delays in signal propagation around the control loop, actuators that do not respond immediately to an external command signal, delays in responding to real time variables (time constants) and sensors acquire data in irregular intervals (feedback delays). Time lags restrict the speed and extent the effects of disturbances, both within the process itself and externally derived. They also impose extra requirements on the controller and needs to infer delays that are not directly observable [1]. Efficiency can be directly related with the capacity to perform the acquisition and processing of inertial data to the space vehicle control system. This efficiency solution must help to strictly comply with the execution time, to ensure proper the spacecraft navigation and control during its flight mission. The critical computer functions should be made as efficient as possible and the processes and threads scheduled by the software should be achievement inside the timing deadline. Acquire information about the software and hardware and study the correct dimension of the computational power of the processor, memory speed and capacity. In addition, how the module (hardware or software) responds to an event has to be known and the resource consumption view is needed.

8.3.7 Failure Tolerance

Fault-tolerance is the capability to build a space system that continue to operate satisfactorily in the presence of faults. A fault-tolerant system must be able to tolerate one or more transient, intermittent, or permanent hardware faults, software and hardware design errors, operator errors, externally disturbances, or physical damage. There are many ways in which data processing may fail—through software and hardware, and whenever possible, spacecraft systems must be capable of tolerating failures [28]. Failure tolerance is achieved primarily via hardware, but inappropriate software can compromise the system's failure tolerance. During the real-time software project, it is necessary to define a strategy to meet the system's required level of failure tolerance. If it is well designed, the software can detect and correct errors in an intelligent way. NASA has established levels of failure tolerance based on two levels of acceptable risk severity: catastrophic hazards must be able to tolerate two control failures and critical hazards must be able to tolerate a single control failure [29].

Examples of software failure are the input and output errors of sensors and actuators. This failure could be tolerated by checking the data range and forcing the software to assume an acceptable value. An example of hardware failure in electronic components is the single-event upset (SEU), an annoying kind of radiation-induced failure. SEUs and their effects can be detected or corrected using some mitigation methods like error detection and correction

(EDAC) codes, watchdog timers, fault rollback, and watchdog processors. According to Siewiorek and Narasimhan [30], the techniques to response to failure can be typically classify as fault confinement (containment), fault detection, diagnosis, reconfiguration, recovery, restart, repair, and reintegration. Rennels [31] describes the evolution of fault-tolerance in spacecraft computers beginning at 1960s, when was developed was the on-board computer for the Orbiting Astronomical Observatory (OAO), using fault masking at the component (transistor) level. After that, the new Jet Propulsion Lab (JPL) approach called Self-Testing-and-Repairing (STAR) computer worked for a 10-year mission to the outer planets. After, the JPL-Voyager computers used pairs dynamic redundancy, in which one computer check the other by exchanging messages and worked in space for more than 20 years.

8.3.8 Maintainability

Maintainability is the capability of the space system can be restored to the operational status following a failure or an improvement, in an easy and fast way. It must be easy for the space computer systems to maintain their subsystems, modules or components during any phase of the mission, whether on the ground or in space. The purpose of maintenance can be repairing a discovered error or allow that a system upgrade to include new features of improvements is made. As an example, one can cite the maintenance performed remotely by NASA on Mars Exploration Rovers Spirit and Opportunity, launched toward Mars in 2003. According to Jet Propulsion Laboratory site information [32], the communications with the Earth is maintained through the Deep Space Network (DSN), an international network of antennas that provide the communication links between the scientists and engineers on Earth to the Mars Exploration Rovers in space and on Mars. Through the DNS, it was possible to detect a problem in the first weeks of the mission that affected the Spirit rover's software, causing it to remain in silence for some time, until the engineers could fix the error. The failure was related to flash memory and it was necessary a software update to fix it. It was also noted that if the rover Opportunity had landed first, it would have had the same problem. In order to improve the maintainability is necessary to implement modularity, readability and complexity management of the code, take into account performance constraints (available memory size, processor load, cycle time, size of the data handled, etc.) to ensure the possibility of being able to change the application, documentation must be update, correct, legible, and in a format to allow analysis and manipulation [33].

8.3.9 Modularity

Modularity is the capability of the system be composed of discrete components such that a change to one component has minimal impact on other components. Modularity is a principle that states that the functionality of a system should be distributed through various nodes so that if a single node is damaged or destroyed, the remaining nodes will continue to function [34]. The partitioning of critical systems in modules provides advantages such as easily maintainability, traceability of the design to code, and allow the distributed software development. Modularity contributes to the verification and validation process and errors detection during the unit, component and integration tests as well maintenance activities. The modularity facilitates the failure isolation, preventing their spread to other modules. The independent development assists the implementation and integration. As an example, a space software configuration item (ICSW) can be divided into software components (CSW), which can be divided into units or modules (USW), which correspond to the tasks to be performed during the preflight and flight phases, in the interaction with the communication interfaces, sensors and actuators, and the transmission of data to the telemetry system.

8.3.10 Portability

Portability is the capability of the space system be transferred from one hardware, software or environment to another. The space projects can be long-term and during its development, there may be situations that require technological changes both to improve the application, and to overcome problems such as the exchange of equipment due to the high dependence on products suppliers. For example, it is desirable that the code can be compiled into an ANSI standard in the space software systems. This will enable the code to be run on different hardware platforms and in any compatible computer system, making only specific adaptations to be transferred from one environment to another.

8.3.11 Reliability

Reliability is the capability to the space system does not fail for a given period of time under specified operating conditions. Space computer systems reliability is dependent on other factors like correct selection of components, correct derating, correct definition of the environmental stresses, restriction of vibration and thermal transfer effects from other subsystems, representative testing, proper manufacturing, and so on [35]. Reliability is calculated using failure rates, and hence the accuracy of the calculations depends on the accuracy and realism of our knowledge of failure mechanisms and modes. For most established electronic parts, failure rates are well known, but the same cannot be said for mechanical, electromechanical, and electrochemical parts or man. The author states that, in modern applications in which computers and their embedded software are often integrated into the system, the reliability of the software must also be considered.

One way to define acceptable reliability levels for space systems is by regulatory authorities and in the case of components, by the manufacturers industries. An example of a space system reliability case history was cited by [28]. The Asteroid Rendezvous (NEAR) spacecraft had a 27-month development time, a 4-year cruise to the asteroid, and spent 1 year in orbit about the asteroid EROS. The spacecraft was successfully landed on EROS in February 2001 after 1 year in orbit. Reliability was maximized by limiting the number of movable and deployable mechanical and electromechanical systems. Determine reliability metrics for system such as Probability of failure on demand (POFOD) or Rate of occurrence of failures (ROCOF). POFOD is the probability that the system will fail when a service request is made. Useful when demands for service are intermittent and relatively infrequent. ROCOF reflects the rate of occurrence of failure in the system. Relevant for systems where the system has to process a large number of similar requests in a short time. Mean time to failure (MTTF) is the reciprocal of ROCOF.

8.3.12 Recoverability

Recoverability is capability to recovery the space system after a failure or after a computer system hacked. In the autonomous embedded systems, that is, that do not require human operators and interact with sensors and actuators, failures with severe consequences are clearly more damaging than those where repair and recovery are simple [21]. Therefore, the embedded computer systems must be able to recover themselves if during the space mission situations where it is not possible to perform the maintenance occur.

As an example, in the execution of a software embedded application during the launcher flight, it is recommended that the function responsible for acquiring the data has a mechanism for recovery if there is a failure that does not allow the Inertial System's data reading, in order to provide the recovery of this information to the control system so that the vehicle is not driven to a wrong trajectory. Recoverability can also be reach through checkpoints adjusted by the system administrator, partition the files across more disks or increasing the system resources available.

8.3.13 Robustness

Robustness is the capability of a space system to cope with errors during execution and cope with erroneous input or abnormal environmental conditions. In addition to physically withstand the environment to which they will be submitted, computer systems must also be able to deal with circumstances outside the nominal values, without causing the loss of critical data that undermine the success or safety of the mission. In case of hardware failure or software errors at run time, the system's critical functions should continue to be executed.

As an example of software robustness assessment, NASA [29] mention fault injection, which is a dynamic-type testing because it must be used in the context of running software following a particular input sequence and internal state profile. In fault forecasting, software fault injection is used to assess the fault tolerance robustness of a piece of software (e.g., an off-the-shelf operating system). Formal techniques, such as fuzzy testing, are essential to showing robustness since this type of testing involves invalid or unexpected inputs. Alternatively, fault injection can be used to test robustness.

8.3.14 Safety

Safety is the capability of the space system is free from those conditions that can cause death, injury, damage to or loss of equipment or property, or damage to the environment. According to [25], the overall objective of the safety program is to ensure that accidents are prevented and all hazards, or threats, to people, the system and the mission are identified, and controlled. Safety attribute is applied to all program phases and embrace ground and flight hardware, software and documentation. They also endeavour to protect people from man-induced hazards. In the case of manned spacecraft, safety is a severe design requirement and compliance must be demonstrated prior to launch.

Hazards can be classified as "catastrophic," "critical," or "marginal" depending on their consequences (loss of life, spacecraft loss, injury, damage, etc.). Also, the most intensive and complete analysis can be done by constructing a safety fault tree. The software safety requirements should be derived from the system safety requirements and should not be analysed separately [15]. In the software space systems, an indicator of criticality for each module defining the level of associated risk, called the safety integrity level should be specified. The most critical modules involve greater strictness in their development process [16]. It must be applied new approaches to expand the approach of traditional safety techniques that analyses only of traditional component failures. The STAMP (System-Theoretic Accident Model and Processes) [1] model also considers design flaws, incomplete or inadequate requirements, dysfunctional interactions among subsystems or components (all of which may be operating exactly as specified), human interactions, and other causes of accidents and incidents.

8.3.15 Security

Security is the capability to assure that all space assets, mainly data inside the system will be protected against malware attacks or unauthorized access. Space systems have as a feature to protect information, due to the strategic interest of obtaining the technology of satellite launch vehicles, currently still dominated by few countries in the world. There should be a strict control in the access to information in these systems, because if a change occurs accidentally or maliciously, this can compromise the success of a mission. Barbacci et al. [36] emphasizes that in government and military applications, the disclosure of information is the primary risk that was to be averted at all costs. As an example of the influence of this attribute, a remote destruction command of a spacecraft launch system must be able to block another command maliciously sent from an unknown source, which seeks to prevent the vehicle from being destroyed, when it violates the flight safety plan. Other attacks in space system can be through hacking a satellite Internet provider (stolen IP address), introducing noise in satellite receiver spectrum, spoof a GPS satellite or even steal sensitive data from ground space system [37]. Security analysis must be performed, identifying the main threats that your system shall be protected, including comprehensive authorization and authentication scheme for each system components (hardware, software, and human), introducing security constraints on who can generate, view, or manipulate the data and consider applying international standards and lessons learned from security community.

8.3.16 Self-description

Self-description is the capability to describe or explain the space software about the context of use. Nowadays, the reuse of technology is common in the course of space programs, that is, many systems or subsystems are reused in subsequent missions, and so require maintenance or adjustments. To minimize the possibility of introducing errors in the project, it is desirable that the computer system to be reused has a description that allows an easy understanding. For example, it is recommended that the code of a software application has comments that explain the operation of its functions, thus facilitating developers to carry out future changes required.

8.3.17 Simplicity

The simplicity is the capability of the space system to became easy and simple to use and maintain its functionalities. Simplicity is an essential aspect for the software used in critical systems, since the more complex the software is the greater the difficulty in assessing its safety [18]. This is a desirable feature in a space software application because functions with simple code have expected operation and are therefore safer than others with difficulties in their understanding and which can produce indeterminate results. Software simplicity is also related to the ease of maintaining its code. For example, a system level fault occurred with Mars Exploration Rover in 2004 put it in a degraded communication state and allowed some unexpected commands [16]. IV&V findings related to the system memory showed that portions of the file system using the system memory was consistently reported to be very complex and modules were reported to have poor testability and poor maintainability. The file system was not the cause of the problem but brought the lack of memory to light and created the task deadlock.

8.3.18 Stability

Stability is the capability to the space system or space software performing their function normally along the time or after performing the changes. Space computer systems require high reliability, and their subsystems and components must continue to perform their functions within the specified operational level without causing the interruption of service provision during the mission, even if the system is operated for an extended period of time. Examples are the satellites that depend upon the performance of solar cell arrays for the production of primary power to support on-board housekeeping systems and payloads throughout their 7–15 years operational lifetime in orbit. The positioning systems of solar panels must have stable operation during the long-term missions, so that the satellite keeps the solar cell arrays towards the sun when going through its trajectory. In 2013, the United Space Alliance, which manages the computers aboard the International Space Station, has announced that the Windows XP computers aboard the ISS have been switched to Linux. They migrated key functions from Windows to Linux because they "needed an operating system that was stable and reliable."

8.3.19 Survivability

Survivability is the capability of a space system to operate without degraded performance if exposed to natural and/or hostile environments. The space systems are designed to operate in an environment with different features from those on Earth, such as extreme gravity, temperature, pressure, vibration, radiation and EMI variations, and so on. Fortescue et al. [25] noted that the different phases in the life of a space system, namely, manufacture, pre-launch, launch and finally space operation, all have their own distinctive features. Although the space systems spend the majority of their lives in space, it is evident that it must survive to the other environments for complete success. Critical systems should continue to provide their essential services even if they suffer accidental or malicious damage. This includes the system being able to: resist to risks and threats, eliminating them or minimizing their negative effects; recognize accidents or attacks to allow a system reaction in case of its occurrence and recovery after the loss or degradation due to an accident or attack [19].

8.3.20 Testability

Testability is the capability to validate the space system through tests, fault detection techniques, and behavior diagnosis A comprehensive spacecraft test program requires the use of several different types of approaches and test facilities. The facilities are required to fulfil the system testing requirements and may include some facilities like clean room, vibration, acoustic, EMC, magnetic, and RF compatibility [28]. In the case of a critical software system, this feature is crucial, especially during the unit test, integration, system and acceptance and validation phases [18]. The real-time software application should be tested as much as its functionality and its performance, ensuring the fulfilment of its functions during the mission within the specified time.

8.3.21 Traceability

Traceability is the capability to track different artefacts of the space system during the development or during the operation process. Particularly important for computer system requirements should linked the requirements to models and to code, enabling the verification and validation through the test cases. This also represents the possibility of mapping the safety requirements in all the system development phases. A Traceability Matrix should be implemented as a document that co-relates any two-baseline documents and artefacts that require any relationship to check its completeness.

8.4 Conclusion

These meta-requirements selected for space systems and discussed in this work can be used as a reference in the requirements engineering phase and should be considered according to the profile of the mission or project, complying standards and regulations or even to improve the quality attributes of the system. Is important to mentioned that this set of meta-requirements is not enough to cover all expectations of the stakeholders. Due to the increase complexity of the space systems today, more and more new nonfunctional requirements must be considered.

Others meta-requirements not presented in this work, but in study are usability, affordability, resilience, reusability and others. This work of identification new meta-requirements for space are (always) on going, as well as their techniques and metrics related.

Acknowledgments

I would like to thank Professor Edgard Yano, from Institute Technological of Aeronautics, ITA and the Technologist Marcos Romani from Institute of Aeronautics and Space, IAE for the contributions to this work during the period from 2007 to 2010.

References

[1] Leveson N. Engineering a Safer World. Systems Thinking Applied to Safety. MIT Press. 2021.

[2] Leveson, N.G. "The Role of Software in Spacecraft Accidents". *AIAA Journal of Spacecraft and Rockets* 41, no. 4 (July 2004): 564-575.

[3] Romani, M. A. S.; Lahoz, C. H. N.; Yano, E. T. Identifying dependability requirements for space software systems. J. Aerosp.Technol. Manag., São José dos Campos, Vol.2, No.3, pp. 287-300, Sep-Dec., 2010.

[4] Ingham M. D., Rasmussen R. D., Bennett M. B., Moncada A. C. 2004. Generating requirements for complex embedded systems using state analysis. 55th International Astronautical Congress, Vancouver.

[5] Hecht, M, Buettner, D. "Software Testing in Space Programs", Crosslink, Volume 6, Number 3 (Fall2005).

[6] Lyons J., Yvan Choquer. Ariane 5: Flight 501 Failure Report By The Inquiry Board Paris , 19 July 199J. 2013.

[7] Mars Climate Orbiter Mishap Investigation Board Phase I Report November 10, 1999.

[8] Report on the Loss of the Mars Polar Lander and Deep Space 2 Missions. JPL Special Review Board. March 2000.

[9] Hjortnaes, K. ESA Software Initiative. Report. May 2003.

[10] Firesmith D.G. "Engineering Safety-Related Requirements for Software-Intensive Systems", Proceedings of the 28th International Conference on Software Engineering, ACM SIGSOFT/IEEE, Shangai, China, pp. 1047-1048, 2006

[11] Romani, M.A.S. "Processo de Análise de Requisitos de Dependabilidade para Software. Espacial". Masters dissertation, Instituto Tecnológico de Aeronáutica, 2007 (in Portuguese).

[12] Lahoz, C.H.N. "Elicere: O processo de elicitação de metas de dependabilidade para sistemas computacionais críticos: estudo de caso aplicado a área espacial." Doctorate thesis, Universidade de São Paulo, São Paulo, 2009 (in Portuguese).

[13] ABNT, Associação Brasileira de Normas Técnicas. Sistemas Espaciais - Gerenciamento do Programa – Parte 2: Garantia do Produto, NBR 14857-2, 2002.

[14] MOD, UK Ministry of Defence. Reliability and Maintainability (R&M) – Part 7 (ARMP -7) NATO R&M Terminology Applicable to ARMP's, Def Stan 00-40, Issue 1, 2003.

[15] ESA, European Cooperation for Space Standardization. ECSS-E-ST-40C rev.1, *Space Engineering - Software*, 2016.

[16] NASA. *Software Safety Guidebook*, NASA-GB-8719.13, 2004a.

[17] Kitchenham, B. and Pfleeger, S.L. "Software Quality: The Elusive Target", IEEE Software, 12-21, January 1996.

[18] Camargo Junior, J.B., Almeida Junior, J.R. and Melnikof, S.S.S. "O Uso de Fatores de Qualidade na Avaliação da Segurança de Software em Sistemas Críticos". *Anais da Conferência Internacional de Tecnologia de Software: Qualidade de Software*, (8), v.1 1997, p. 181-195.

[19] Firesmith, D.G. "Common Concepts Underlying Safety, Security, and Survivability Engineering", Technical Notes 033, Software Engineering Institute/Carnegie Mellon University, Pittsburgh, USA, 70 p., 2003.

[20] Rus, I.; Komi-Sirvio, S. and Costa, P. "Software Dependability Properties: A Survey of Definitions, Measures and Techniques". High Dependability Computing Program (HDCP). Fraunhofer Center for Experimental Software Engineering, Maryland, Technical Report 03-110, 2003.

[21] Sommerville, I. "Software Engineering", 7th Ed. Addison-Wesley, Glasgow, UK, 2004, 784 p.

[22] Bass, L.; Clements, P.; Kazman, R. Software architecture in practice. 3. ed. Boston, MA: Pearson Education, 2013.

[23] Ericson, C. A. Hazard analysis techniques for system safety. 2. ed. Hoboken, New Jersey: John Wiley & Sons, 2016.

[24] Avizienis, A.; Laprie, J.-C.; Randell, B.; Landwehr, C. Basic concepts and taxonomy of dependable and secure computing. IEEE transactions on dependable and secure computing, IEEE, v. 1, n. 1, p. 11{33, 2004.

[25] Fortescue, P., Stark, J. and Swinerd, G. "Spacecraft Systems Engineering", 3rd Ed. John Wiley & Sons, London, UK, 2003, 678 p.

[26] Zowghi, D. The Three Cs of Requirements: Consistency, Completeness, and Correctnes. Proceedings of 8th International Workshop on Requirements Engineering: Foundation for Software Quality. 2020.

[27] Lutz, R.R. "Analyzing Software Requirements Errors in Safety-Critical, Embedded Systems". Ames: Iowa State University of Science and Technology – Department of Computer Science, Technical Report 92-27, 1992.

[28] Pisacane, V.L. "Fundamentals of Space Systems", 2nd Ed. Oxford University Press, New York, USA, 2005, 828 p.

[29] NASA. *Software Fault Tolerance: A Tutorial*, Technical Memorandum NASA/TM-2000-210616, Langley Research Center, Hampton, USA, 2000, 66 p.

[30] Siewiorek, D.; Narasimhan, P. Fault-tolerant Architectures for Space and Avionics Applications. Electrical and Computer Engineering Department. Carnegie Mellon University. Published 2005.

[31] D. A. Rennels, Fault-Tolerant Computing, Encyclopedia of Computer Science, ed., Anthony Ralston, Edwin Reilly, and David Hemmendinger, 1999 (International Thomson Publishing – Europe).

[32] Jet Propulsion Laboratory (JPL). *Mars Exploration Rover Mission – Communications with Earth.*

[33] Boulanger, Jean-Louis.Certifiable Software Applications 1: Main Processes. ISTE Press – Elsevier. 2016.

[34] Jackson, Scott. Principles for Resilient Design - A Guide for Understanding and Implementation. In IRGC Resource Guide on Resilience, edited by I. Linkov. University of Lausanne, Switzerland: International Risk Governance Council (IRGC) 2016.

[35] Fortescue, P., Stark, J. and Swinerd, G. "Spacecraft Systems Engineering", 3rd Ed. John Wiley & Sons, London, UK, 2003, 678 p.

[36] Barbacci, M., Klein, M.H., Longstaff, T.A. and Weinstock, C.B. "Quality Attributes", Technical Report CMU/SEI-95-TR-021, Software Engineering Institute/Carnegie Mellon University, Pittsburgh, USA, 1995, 56 p.

[37] Falco, G. Cybersecuriy Principles for Space Systems. Journal for Aerospace Information Systems. Dec 2018.

9

The Role of Requirements Engineering in Safety Cases

Camilo Almendra[1] and Carla Silva[2]

[1]Federal University of Ceará (UFC), Brazil
[2]Federal University of Pernambuco (UFPE), Brazil
E-mail: camilo.almendra@ufc.br; ctlls@cin.ufpe.br

Abstract

Safety cases (SCs) are a novel approach to turn safety certification documentation more structured and comprehensible. This approach aims to foster communication and discussion over the safety assurance of a system using arguments structures. Safety requirements are building blocks of SCs which are mapped into the safety arguments based on Claim-Argument-Evidence structures. Requirements can be made into safety arguments in many ways, depending on the type of requirement and the argumentation pattern. We discuss examples of requirements presence in SCs and also their participation in safety argumentation patterns. SC development should be performed alongside the system construction, so safety concerns are addressed since the early phases of system development. We discuss how the integration of SCs development and the Requirements Engineering activities could occur.

9.1 Introduction

In this chapter, we present an overview of the concept of Safety Cases (SCs), notations and development methods for SCs, as well as we discuss the role and influence of Requirements Engineering (RE) in SCs development.

9.2 Safety Cases

In the middle 1990s, regulatory agencies and industry faced many limitations toward safety certification, especially in software systems [1]. For years, the regulations mainly focused on establishing prescriptive, obligatory practices and processes that vendors should perform [2]. However, the justification that a software-intensive system is ready for use is considered a complex social-technical process, as technologies, environment, and practices rapidly evolve in the software field [1]. Then, some regulations moved toward adopting a descriptive approach to define goals, requirements and constraints that systems should satisfy. In this new paradigm, certification authority agents scrutinize a system and its documentation, looking for evidence that safety goals were satisfied by system requirements and design.

9.2.1 Definition

Safety case is a novel approach to communicate and argue over the safety of a system, offering more flexibility to justify software requirements and design decisions. The purpose of an SC is to present *"a clear, comprehensive and defensible argument that a system is acceptably safe to operate in a particular context"* [4]. They are also referred to as "safety assurance cases," as the idea of developing cases to demonstrate the fitness to use of systems has spread to other concerns such as security and dependability [5].

SCs are an amalgamation of information from standard requirements, product requirements, risk and hazard analysis results, design decisions and rationale, validation and verification results, and process management records [1]. Textual forms were the original approach to present safety argumentation, and SC reports comprised hundreds of pages of unstructured text.

The need for a more formal structure has led to the development of textual and graphical notations to specify SCs toward safety arguments. The *claim-based* approaches pioneered and influenced research and practice. In this kind of approach, the structure of safety arguments comprises three main types of element [3]:

- Claims: high-level goals, subgoals, or constraints addressed during the system development.
- Arguments: reasoning steps used to show how pieces of evidence support the claims.
- Evidence: any piece of information that substantiates some claim or argument generated during design, implementation, and verification activities.

Often, a top claim needs to be decomposed into subclaims, creating a hierarchy of claims that go from top claims down to evidence items. Figure 9.1 illustrates the relationship between claims, arguments, and evidence in a safety argument structure.

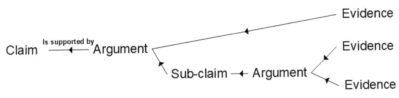

Figure 9.1   Safety argument structure (adapted from [3]).

The Adelard Safety Case Development (ASCAD) methodology and notation [6], popularly referred to as Claim-Argument-Evidence (CAE) was the first approach to organize safety arguments using claim-based frameworks [14]. Another graphical claim-based approach is the Goal Structuring Notation (GSN) [7]. The underlying metamodel of GSN and CAE are equivalent in their capability to express claim-based arguments, even though they comprise different types of concepts and relationships. GSN has gained more traction in the academy and has the support of a community of researchers and practitioners [5].

The first standard to explicitly require SCs for a software was the Interim MoD Def Stan 00-55, whose current version is the Def Stan 00-56, a regulation that covers the UK's defence systems contracts [12]. Examples of regulations that require SCs are the standards ISO 26262 for road vehicles, EN 50129 for railway applications, and FDA guidance for infusion pumps [5].

The standard ISO/IEC 15026-2 ("*Systems and software assurance— Assurance case*") aims to "*improve the consistency and comparability of assurance cases and to facilitate stakeholder communications, engineering decisions and other uses of assurance cases*" [8]. It defines common terminology, minimum needs related to structure and contents of assurance cases, and a life cycle for all types of assurance cases, including SCs. At the same time, there are no restrictions upon notation and tools [8].

9.2.2 Example

Figure 9.2 illustrates the GSN notation and a safety argument example. This example demonstrates a partially developed safety argument that already has

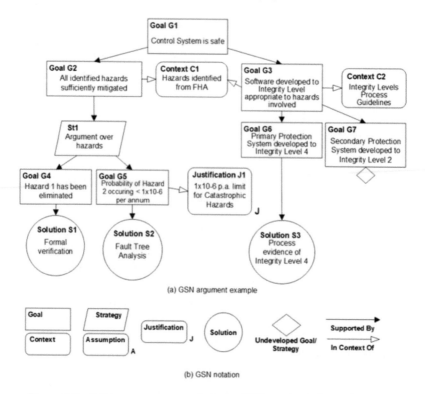

(a) GSN argument example

(b) GSN notation

Figure 9.2   Safety argument example in the GSN notation (adapted from [9]).

concrete evidence traced to some goals, but how to accomplish others is still undefined.

A short explanation follows to illustrate the structure and content of an SC. In Figure 9.2a, the top Goal *G1* decomposes into subgoals *G2* and *G3*. Such decomposition means that the fulfilment of both subgoals implies the fulfilment of the top goal. The subgoals are in the context of a Functional Hazard Analysis (FHA) represented as the context *C1*; such analysis found a list of hazards associated with the Control System. The Context *C2* represents the integrity levels process guidelines, and it affects Goal *G3*. This means that G3 has to be fulfilled in compliance with these guidelines. Strategy *St1* addresses Goal *G2*, arguing over the elimination or acceptable mitigation of each hazard identified.

Then, the next level of decomposition goes down to create separate subgoals (*G4* and *G5*) that decompose Strategy *St1*, each one addressing a single hazard. Goal *G4* states the elimination of *Hazard 1* from the system, supported by formal verification (Solution *S1*). Goal G5 states the mitigation of *Hazard 2*, stating that its probability of occurrence is smaller than the specified limit. Justification *J1* holds an explanation for this limit. The evidence Solution *S2* that supports Goal G5 is the output of a Fault Tree Analysis. Goal *G3* decomposes into subgoals *G6* and *G7*. The diamond symbol below *G7* indicates that it is still undeveloped. Goal *G6* is supported by process evidence appropriated for the required integrity level (Solution *S3*). This short example shows how diverse and interconnected is the content of SCs.

Much research was devoted to improving SCs development by supporting graphical notations [5] and tool development [10]. However, industry practice still inclines toward the adoption of textual templates and structured text [5]. Textual SCs can take many different forms, from normal prose to argument outlines. Adopting a controlled textual notation to express safety arguments brings similar gains of graphical approaches by having an underlying syntactic metamodel. Users need to explicitly indicate which elements of the argument they are referring to and indicate the relationships between these elements. Regardless of the visual presentation, the underlying metamodel guides the development of SCs.

The Object Management Group (OMG) developed the Structured Assurance Case Metamodel (SACM) to promote standardization and interoperability between various assurance case notations [13]. SACM intends to leverage the development of SCs, moving away from manual approaches and unstructured textual forms to model-based approaches that support automation of construction and assessment of safety arguments.

9.2.3 Development

The organization of claims that must be addressed, their decomposition into arguments, and the provision of evidence to back up claims/arguments comprise the fundamental objective of Safety Case Development (SCD).

Seminal approaches for SCD were proposed in the late 1990s and beginning of the 2000s, such as the *Safety Case Methodology* by Bishop and Bloomfield [14], the *Six-steps Goal Structure Method* by Kelly [4] and the *Conceptual Framework and Patterns Catalogue* by Weaver [9]. We briefly discuss these approaches below, highlighting their seminal contribution (Figure 9.3).

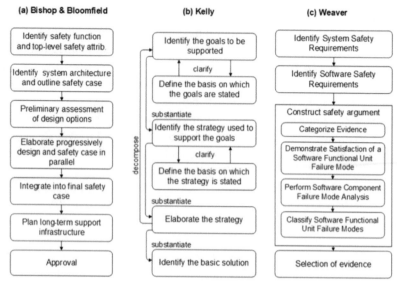

Figure 9.3   Safety case development methodologies (adapted from [4, 9, 14]).

The work of Bishop and Bloomfield [14] was in the context of developing the ASCAD (Adelard Safety Case Development) methodology and CAE notation. Their methodology includes guidance for classifying types of claims, types of argument, possible sources of evidence, and strategy for layering SCs (Figure 9.4a). Since the beginning of research and practice toward SCs, there was a recommendation for integration with system/software design and development processes. Bishop and Bloomfield noted that safety assessment is a significant development factor. Hence, designers should consider the cost and feasibility of SCD when designing a safety-critical system [14]. The evaluation of a candidate design solution should consider how much effort would take to create the required body of evidence for assessing its fitness for use.

Kelly [4] provided a methodology and a graphical notation for building safety arguments. His work focuses on four main problems toward SCD: how to present clear safety arguments, how to support incremental development, how to manage them throughout their life cycle, and how to support reuse. Kelly proposed a methodology (Figure 9.4b) and two extensions of the GSN notation to enhance it for supporting abstraction/modularity and a structure for SC pattern documentation.

Weaver [9] proposed a conceptual framework that systematizes evidence required for building software safety arguments. This work presented a method (Figure 9.4c) and a set of SC patterns described in GSN, which serves as basis to build arguments using the framework.

In the early years of SCD, the traditional approach was to elaborate the case in the late phases of system development, which led to some problems. First, it increases rework caused by late-discovered safety issues. Also, it forced argumentation over a given design, with limited space for exploring alternatives. Additionally, developers tend to lose decision rationale as the project timeline may be spread over months and with personnel changes.

The early adopters and designers of SCD methodologies also recommend integrating SCD into system development processes, but this is still a challenge [11]. We discuss the integration of SCD and RE in the remainder of this chapter.

9.3 Requirements Artefacts and Safety Cases

Safety cases are the pinnacle of traceability, as they are artifacts that interrelate requirements, design and implementation information, and comprise rationale on the many decisions taken during system development [19]. Thus, elements of SCs could refer to many types of project management information and requirements are essential building blocks of SCs.

In this section, we discuss the relationship between requirements and argument elements. We illustrate the presence of requirements in SCs using examples based on the GSN notation.

9.3.1 Safety Requirements

The complete set of requirements for a safety-critical system comprises different requirements at various levels of detail. The cross-cutting nature of safety concerns reflects in the classification of safety requirements. The Safe-RE taxonomy provides a common vocabulary to support the management of safety-related concepts for various industry standards [17]. In the taxonomy, a *Safety Requirement (SR)* is typically a quality criterion, combined with a minimum or maximum required threshold and some quality measure. A *Safety-Significant Requirement (SSR)* is a functional, data, interface, or nonsafety quality requirement relevant to the achievement of the safety requirements. So, a SSR can lead to hazards and accidents when not implemented correctly. *Pure Safety Requirements (PSR)* describe the system's expected behaviour in a safe state. Finally, *Safety Constraint (SC)* is an architecture or design constraint requiring the use of a specific safety mechanism or safeguards. The interrelation between these concepts is depicted in Figure 9.4.

Requirements contribute to SCs in many ways. This section provides examples of requirements representation in SCs elements gathered from Hatcliff et al. [15] and Weinstock and Goodenough [16]. We follow the Safe-RE taxonomy [17] to classify the requirements found in the examples.

Requirements as Goals. A straightforward way to map requirements into safety arguments is to state them as Goals. Requirements in SCs may refer to various goals such as reliability, usability, security, safety, functional correctness, accuracy, time response, and obustness to overload.

High-level safety requirements fit well as Goal elements at the top of the argument hierarchy. All seminal approaches discussed previously begin the SCD by identifying and stating high-level requirements as top goals. Then, they decompose into more granular functional or nonfunctional requirements to the point that the argument reaches design and implementation decisions. The process of argument decomposition continues down to the pieces of evidence.

For example, Figure 9.5 shows the top goals stated in the SC for a generic Patient Controlled Analgesia (PCA) Pump [15]. Goal *G1* is a top safety requirement of the system, which decomposes into subgoals *G2* and *G3*. The argumentation strategy aims to demonstrate effectiveness and safety in separation. In the first subgoal *G1*, the argumentation focuses on effectiveness. In the other subgoal *G2*, the focus is on requirements devised to mitigate or eliminate hazards.

Figure 9.6 shows subgoal *G2.1.1* that represents a system requirement. This requirement should be implemented to mitigate the risk of underdose or

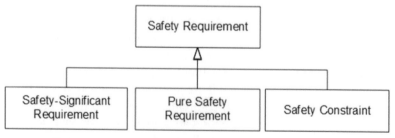

Figure 9.4   Type of safety requirements (adapted from [17]).

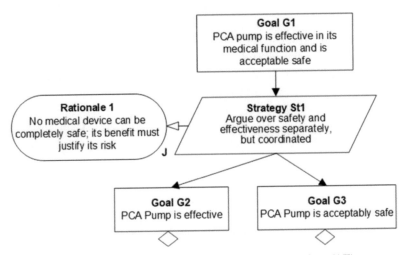

Figure 9.5 Top-level goals in a safety case (adapted from [15]).

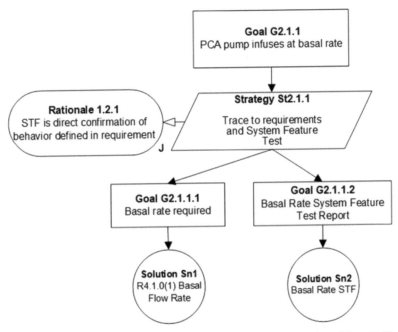

Figure 9.6 Goal decomposition into arguments and evidence (adapted from [15]).

overdose on patients. Thus, it is a *SSR* in the Safe-RE taxonomy. Examples from Figures 9.5 and 9.6 are from the same SC. To fulfil Goal *G2*, the argument focuses on demonstrating the effectiveness of the implementation of subgoals, including Goal *G2.1.1*.

Let move to another medical device example, a Generic Infusion Pump [16]. In this example, illustrated in Figure 9.7, we also find the mapping of requirements into Goals. Goal *G.1.1* is a *PSR* that supports the mitigation of the hazard related to battery exhaustion (the parent Goal *G1*). Then, the PSR is decomposed into a low-level requirement, stated as Goal *G1.1.1* which is an example of *SSR*. It is worth noting that new hazards arise from this requirement decision, which triggers a new round of safety and requirements analysis (Strategy *St1.1.1*).

In the Safe-RE taxonomy, *Safety Constraints* encompass the nonfunctional requirements (NFR) that are related to safety. The cross-cutting nature of NFRs affects the way they are linked and managed during system development, as a *Safety Constraint* may impose restrictions in many functional requirements. Therefore, constraints can appear as Contexts or as Goals in the SCs.

Figure 9.8 shows Goal *G2* which represents the hazard of entry error caused by keypad design. Subgoals *G2.1–G2.4* represent safety constraints devised to address this hazard.

Requirements as Strategy. Strategies provide elements for reviewers (e.g., certification authority) to assess the trustworthiness of the argument step. The

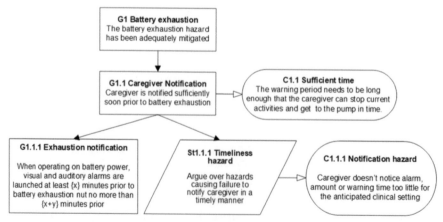

Figure 9.7   Goal decomposition and derivation of new hazards (adapted from [16]).

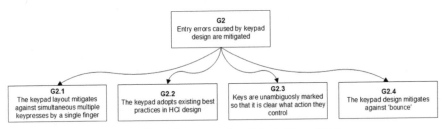

Figure 9.8 Safety constraints (adapted from [16]).

rationale behind a requirements decomposition may not be apparent or trivial, thus requiring analysts to record an explanation. Strategy elements represent the rationale that supports the transition from a parent goal to its subgoals.

In Figure 9.5, Strategy *St3.1* states there is traceability information linking to requirements specification and test specification. Figure 9.8 shows an example of some requirements stated together in an argument. Strategy *St3.1* argues that the combination of two requirements mitigates the hazard in the parent Goal.

Requirements as Context, Justification, or Assumption. Contextual information (Context, Assumption and Justification elements of GSN— Figure 9.2b) can be attached to goal structures for facilitating the

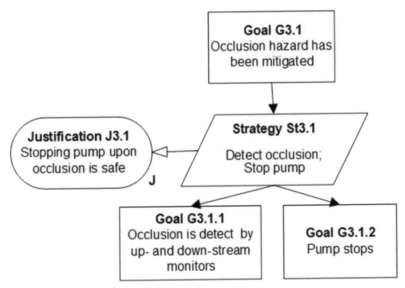

Figure 9.9 Requirements for hazard mitigation and the rationale (adapted from [15]).

understanding of the argumentation approach. The use of Justification and Assumption provides a more specific semantic to the SC. However, the registering of requirements rationale as Contexts is valid. These contextual elements provide essential information to improve confidence in the arguments.

Requirements rationale often appears as Justifications. In Figure 9.6, the Justification *Rationale 1.2.1* states that a successful pass in the System Feature Test (SFT) should confirm the expected behavior.

In Figure 9.5, the Justification *Rationale 1* states an explanation for the decision to separate the safety argumentation for the system.

In Figure 9.7, it is included the requirement rationale as the Context *C1.1*. In Figure 9.9, the rationale is provided as a Justification *J3.1*. For each requirement, the argument is further decomposed until the evidence items.

In Figure 9.10, the Assumption *A2.2* relates to the requirements process, assuming that the requirements specification document holds the entire intended behaviour for the system.

Requirements as Solutions. Solutions are concrete evidence items that support goals. Requirements may appear as Solutions, either as single requirements items or as a reference to an entire specification document. Figure 9.6 shows a requirement item (Solution *Sn1*) serving as evidence for a Goal.

Solutions may be related to the product's behavior or to the process used to develop it. For example, a product's behavior evidence is achieved by the execution of verification and validation activities. In the case of simulation,

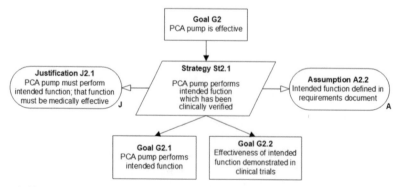

Figure 9.10   Justification and assumption related to PCA pump performs intended function (adapted from [15]).

the results of requirements models validation also serve as evidence that the product's design met expected quality properties. Process-related evidence items, requirements specification review records, and requirements-based hazard analysis are types of solutions that increase confidence in the development process.

9.3.2 Argumentation patterns

The backbone of safety argumentation shall demonstrate how the system design and implementation address hazards and risks. In software-intensive safety-critical systems, the software requirements are essential elements in the safety argumentation.

Argument development is still a challenging task for teams [11, 13], and there are many ways to organize an SC. Argument patterns catalogues have been proposed [4, 9, 18] to leverage the process of building arguments. We discuss some argument patterns in which requirements are central participants.

Arguing over functions. A classic "divide and conquer" approach to deal with systems complexity is to decompose high-level requirements into smaller, manageable, and verifiable functions.

The Functional Decomposition Pattern structures the safety argumentation toward the safety of each system function [4]. In Figure 9.10, Goal *G1* represents the main claim for the safety of a generic system (or subsystem). Strategy *St1* presents the functional decomposition approach to support the top goal. Context *C1* shall bring the list (or a reference to it) of all safety-related requirements ("functions") of the system. The argument structure then iterates over all known requirements (Goal *G2* and the relationship *SupportedBy[n]* represents a template to state subgoals for each known requirement). Then, each subgoal is further decomposed down to evidence items. Finally, a subgoal *G3* states that there are no hazardous interactions between the requirements.

Arguing over contribution to hazards. Software, as well as any aspect of systems design, may contribute to hazardous scenarios. Software failure is one type of cause of hazards [17], and the SC needs to explain how its contribution was minimized or eliminated by design.

The Component Contributions to System Hazards Pattern structures the safety argumentation toward identifying hazards and assessing the associated risks [9]. In Figure 9.11, the Strategy *ArgSWHWOther* represents the

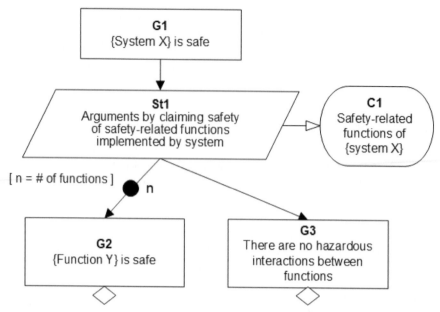

Figure 9.11   Functional decomposition pattern (adapted from [4]).

decomposition of the System Level Hazards across the Hardware, Software, and Other Parts of the system. The Goal *SWContribAccept* claims that all software contributions are acceptably safe. We highlight the participants *SWDef* and *SWContrib*. They are both Contexts elements; the former presents a description or model of the system software, and the latter the safety requirements which are related to the software. This pattern puts hazards as central participants, which are complemented with design and requirements as contextual information.

Arguing over a generic design. The use of software as building blocks of critical systems provides much flexibility in systems design. However, a generic system model may provide a known, easy to assess structure in which vendors could organize the safety argumentation.

The Control System Architecture Breakdown Pattern structures the arguments to support high-level goals toward the decomposition into a generic control system model [4]. The application of this pattern depends on whether the system's architecture under assessment fits the generic model. Figure 9.13 shows Strategy *S1* as the link to subgoals that address separately each safety requirement (*C1*) related to sensors (*G2*), control logic (*G3*),

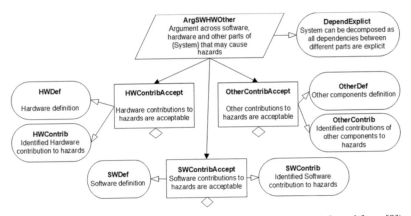

Figure 9.12 Component contributions to system hazards pattern (adapted from [9]).

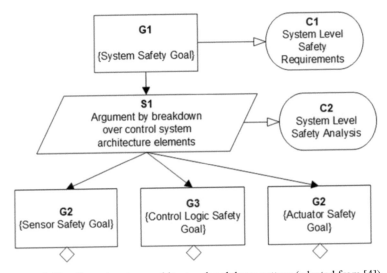

Figure 9.13 Control system architecture breakdown pattern (adapted from [4]).

and actuators (*G4*). Context *C2* holds the output of the safety analysis that identifies how low-level elements of the control system architecture contribute to hazards.

Arguing over multilevel design. There are many approaches to tackle systems complexity, such as the successive refinements of requirements and architecture. In this type of approach, each tier (or level) of requirements specification shall drive good design decisions in the next tier of development.

The Software Safety Requirements Identification Pattern structures the arguments to warrant that safety requirements from a development tier have been adequately addressed at the next tier [18]. This pattern can be applied whenever it is possible to appropriately identify levels of abstraction/modularization in the system design (e.g., packages and class design levels).

Like previously discussed patterns, we again see a tight linking between design and requirements in the organization of safety arguments. This pattern highlights that at a given tier of development, either design decisions mitigate the safety requirements or another step of decomposition derives additional safety requirements addressed in the next tier.

Figure 9.14 shows two top elements: Goal *SSRidentify* and Strategy *SSRidentify*. The goal identifies the tier of development addressed by the following argumentation. The strategy holds that the safety of a tier of development can be demonstrated by: (i) claiming that design decisions in a current tier satisfy the safety requirements of a previous tier (Goal *DesignDecisions*) and (ii) guaranteeing that prior tier safety requirements not addressed are adequately captured by additional safety requirements for the current tier (Goal *SSRcapture*). These two goals are in the context of the design specification for the current tier (Context *tierNDesign*).

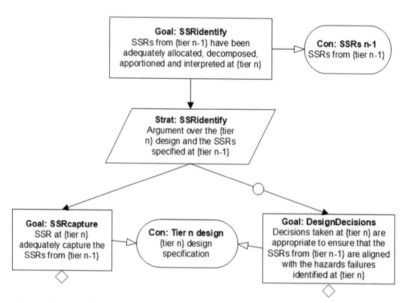

Figure 9.14   Software safety requirement identification pattern (adapted from [18]).

9.4 Safety Case Development and Requirements Processes

As modern development processes rely more and more on exploratory or iterative-incremental approaches, it is challenging to disseminate knowledge about new or derived safety requirements throughout the development processes. Practitioners could better approach SCD as an evolutionary activity integrated with the system development life cycle. The incremental development of SCs fosters early discussion of safety concerns, enables the system design to be reviewed from a safety assurance view, and bridges the expertise of various stakeholders involved in safety-critical systems development [20].

In this section, we discuss some aspects of how RE processes and SCD could better work together.

9.4.1 Joint development

RE activities could have integration points in the SDLC to hook up SCD activities. In RE activities, requirements are hardly the only focus. Stakeholders, users, design, constraints, project management, and various concerns permeate the discussion toward requirements. Similarly, SCD also requires multiperspective thinking.

The first cycles of elicitation and analysis may produce varying versions of the initial system design and requirements specification. For critical systems, also preliminary hazard analysis is produced. Once an initial set of features, high-level design and hazard analysis is available, SCD can start. Indeed, the early production of an SC can help find open safety issues and identify evidence needs that affect project planning. A preliminary SC also provides a place to record rationale information, reducing future efforts to recall it.

Some approaches have been proposed to integrated SCD into agile development. The works of Stålhane and Myklebust [21] and Cleland-Huang and Vierhauser [22] extended the Scrum framework[1] to propose the construction of SCs during or after the sprints. During the sprint, planning preliminary discussion on safety arguments and evidence is carried on; after the sprint, the SC is reviewed by the team and stakeholders alongside the other artefacts produced in the sprint. Thus, in each iteration, the SC is extended

---

[1] Scrum Guide—https://scrumguides.org

Table 9.1 Benefits of SCD alongside RE activities.

RE activity	Purposes for RE	Benefits of SCD
Release Planning	Decomposition of features into requirements and identification of hazards and safety requirements.	Production of a preliminary SC helps the early safety assessment of requirements decomposition.
Backlog Grooming	Revision of whole product planning and refinement of requirements for future iterations.	SCs provide a snapshot view of the safety assurance strategy, highlighting points that need further discussion.
Change Management	Analysis of the impact of a change request.	This activity requires support to "what-if" analysis and an SC provides a way to analyze the impact of changes in safety assurance.
Iteration Planning	Discussion toward the scope, acceptance criteria and decomposition of requirements.	SCs can be used to recall the safety evidence needs for the requirements selected for development.
Iteration Review	Validation of requirements implementation and quality of deliverables.	SCs updated after the end of iteration provide stakeholders with an updated status of assurance of the system.

with the claims, arguments and evidence to stay aligned with the project progress.

This kind of approach fits well with the agile culture. Up-front analysis and design should be "just enough" and successive refinements happen throughout the development. Table 9.1 presents some benefits of developing SC during some RE activities. We used the Scrum framework as an illustration of requirements-related activities.

9.4.2 Traceability

Certification of a system requires evidence information from all development disciplines. To meet safety assurance goals, traceability must be supported. Several problems related to traceability could delay the certification of critical products, such as lack of a traceability information model, uneven granularity of tracing links, and missing or redundant links [23].

Requirements traceability activities executed alongside development play a role in feeding the SCD. In critical system development, risk and hazard analysis artefacts bring an additional dimension for software traceability. SC construction demands from the development team planning and disciplined execution of requirements traceability tasks.

The avoidance of duplicated information is challenging in SCD. SCs are a purpose-specific document to be used mainly in the context of certification procedures. Naturally, SCs elements reference or replicate information usually registered elsewhere in the project's management systems and artefacts repositories. Such duplication of information may increase cost and inconsistencies as the project scales. It is essential to provide means for partially or fully automated generation of SCs from project information systems and artefacts.

9.5 Conclusions

In this chapter, we presented an overview of the interrelationships between RE and SC development. Safety cases comprise information from a variety of activities in the development process, including RE. Although there is extensive literature on SC construction and assessment, practitioners still find challenging the selection and organization of project information into safety arguments. We sought to illustrate the presence of requirements-related information in SCs using examples and patterns from the literature. Requirements happen to appear in all parts of SCs, depending on the perspective of the argumentation. In feature-driven argumentations, safety requirements are the high-level goals or claims that need to be addressed in system design. On the other way, in hazards-driven argumentations, requirements are more likely to appear as supporting arguments and evidence to corroborate the proper mitigation of hazardous scenarios.

The integration of SC development into systems engineering processes is a recommended practice yet challenging for practitioners. We discuss the joint development of SC and RE, highlighting RE practices that would benefit from the integration. During RE activities, requirements are hardly the only aspect discussed. Requirements professionals ordinarily look at aspects beyond requirements, such as design, testing and implementation. Also, the involvement of multiple stakeholders is expected in RE activities, thus turning them into suitable moments to trigger SC development.

Finally, we discuss the requirements traceability as a fundamental practice to support safety case construction. Safety-critical systems development demands mature traceability practices from teams as safety cases are built upon the rationale that binds together requirements, design, and implementations decisions.

References

[1] Bloomfield, R., & Bishop, P. (2010). Safety and assurance cases: Past, present and possible future - An adelard perspective. In C. Dale & T. Anderson (Eds.), Making Systems Safer - Proceedings of the 18th Safety-Critical Systems Symposium, SSS 2010 (pp. 51–67). https://doi.org/10.1007/978-1-84996-086-1_4

[2] Hawkins, R., Habli, I., Kelly, T., & McDermid, J. (2013). Assurance cases and prescriptive software safety certification: A comparative study. Safety Science, 59, 55–71. https://doi.org/10.1016/j.ssci.2013.04.007

[3] Rushby, J. (2015). The Interpretation and Evaluation of Assurance Cases. SRI International, Menlo Park, CA, USA, (July), 135.

[4] Kelly, T. P. (1998). "Arguing Safety – A Systematic Approach to Managing Safety Cases". Doctoral dissertation, University of York.

[5] Rinehart, D. J., Knight, J. C., & Rowanhill, J. (2017). Understanding What It Means for Assurance Cases to "Work."

[6] Adelard: Claims, Arguments and Evidence (CAE). http://www.adelard.com/asce/choosing-asce/cae.html, Accessed: 2019-05-01

[7] Goal Structuring Notation Community Standard (Version 2), http://www.goalstructuringnotation.info/, Accessed: 2019-05-01.

[8] ISO/IEC. (2011). ISO/IEC 15026-2:2011 Systems and Software Engineering — Systems and Software Assurance — Part 2: Assurance Case.

[9] Weaver, R. A. (2003). The safety of software: Constructing and assuring arguments. Doctoral dissertation, University of York.

[10] Maksimov, M., Fung, N. L. S., Kokaly, S., & Chechik, M. (2018). Two Decades of Assurance Case Tools: A Survey. In B. Gallina, A. Skavhaug, E. Schoitsch, & F. Bitsch (Eds.), Computer Safety, Reliability, and Security (pp. 49–59). https://doi.org/10.1007/978-3-319-99229-7_6

[11] Nair, S., De La Vara, J. L., Sabetzadeh, M., & Falessi, D. (2015). Evidence management for compliance of critical systems with safety standards: A survey on the state of practice. Information and Software Technology, 60, 1–15. https://doi.org/10.1016/j.infsof.2014.12.002

[12] Defence Standard 00-056 Part 2 Safety Management Requirements for Defence Systems Part 2 : Guidance on Establishing a Means of Complying with Part 1. (2017).

[13] De La Vara, J. L., Ruiz, A., & Espinoza, H. (2018). Recent Advances towards the Industrial Application of Model-Driven Engineering

for Assurance of Safety-Critical Systems. Proceedings of the 6th International Conference on Model-Driven Engineering and Software Development, (Modelsward), 632–641. https://doi.org/10.5220/000673 3906320641

[14] Bishop, P., & Bloomfield, R. (2000). A Methodology for Safety Case Development. Safety and Reliability, 20(1), 34–42. https://doi.org/10.1 080/09617353.2000.11690698

[15] Hatcliff, J., Larson, B., Carpenter, T., Jones, P., Zhang, Y., & Jorgens, J. (2019). The Open PCA Pump Project: An Exemplar Open Source Medical Device as a Community Resource. ACM SIGBED Review, 16(2), 8–13. https://doi.org/10.1145/3357495.3357496

[16] Weinstock, C. B., & Goodenough, J. B. (2009). Towards an Assurance Case Practice for Medical Devices. Retrieved from https://resources.se i.cmu.edu/library/asset-view.cfm?assetid=8999

[17] Vilela, J., Castro, J., Martins, L. E. G., & Gorschek, T. (2018). Safe-RE: a Safety Requirements Metamodel Based on Industry Safety Standards. SBES, 196–201. https://doi.org/10.1145/3266237.3266242

[18] Hawkins, R., & Kelly, T. (2013). A software safety argument pattern catalogue. The University of York, York.

[19] Almendra, C., & Silva, C. (2020). Managing Assurance Information: A Solution Based on Issue Tracking Systems. Proceedings of the 34th Brazilian Symposium on Software Engineering, 580–585. https://doi.or g/10.1145/3422392.3422454

[20] Almendra, C., Silva, C., & Vilela, J. (2020). Incremental Development of Safety Cases: a Mapping Study. Proceedings of the 34th Brazilian Symposium on Software Engineering, 538–547. https://doi.org/10.114 5/3422392.3422398

[21] Stålhane, T., & Myklebust, T. (2016). The agile safety case. Lecture Notes in Computer Science (Including Subseries Lecture Notes in Artificial Intelligence and Lecture Notes in Bioinformatics), 9923 LNCS, 5–16. https://doi.org/10.1007/978-3-319-45480-1_1

[22] Cleland-Huang, J., & Vierhauser, M. (2018). Discovering, Analyzing, and Managing Safety Stories in Agile Projects. 2018 IEEE 26th International Requirements Engineering Conference (RE), 262–273. https://doi.org/10.1109/RE.2018.00034

[23] Mäder, P., Jones, P. L., Zhang, Y., & Cleland-Huang, J. (2013). Strategic Traceability for Safety-Critical Projects. IEEE Software, 30(3), 58–66.

10

Safety and Security Requirements Working Together

C. Santos

Vaterstetten, Germany
E-mail: cristiano.constantino.santos@gmail.com

Abstract

Currently, critical systems are making use of connected technologies and devices, being the target of cybersecurity attacks, through vulnerabilities exploitation, in similar ways to conventional systems. The relationship between safety and security domains represents a crucial step to avoid cybersecurity threats in a critical system. The security activities should be integrated into a regular safety process and many security methods shall be applied, including security requirements definition and verification, to identify and avoid possible threats that compromise the system safety assurance.

Keywords: Security, security requirements, safety, threat conditions, threat scenarios, critical system, security measures, security risk assessment, safety assessment.

10.1 Introduction

Someone take over a huge commercial aircraft and nobody from the crew can do something because they are not trained for that situation or they do not even know who did such an impossible thing? Unfortunately, this is not an action or sci-fi movie; this is real life.

Currently, the use of notebooks, smartphones, and Wi-Fi connected devices are a reality in the word as well as the use of Wi-Fi connection is a necessary feature in the aerospace industry. Besides that, increasingly, the aircraft are using connected features, such as GPS and ACARS, and software updates using internet connection. The scenario is even worse if we consider the new type of aircraft intended to be certified, the VTOLs, which are planned to flight autonomous in the future. In order words, any computer that is connected to a network is vulnerable to a malware infection. This combination becomes a perfect scenario for the most common terrorists in the world, the hackers.

Aware of this modern virtual threat, the aeronautical industry together with the certification authorities designed specific security standards to handle the threat of intentional attacks on aircraft safety. The security process should be integrated into the aircraft process at the development level, providing outputs such as security requirements, and using inputs such as system safety assessment, to the aircraft development processes.

10.2 Approaching Safety and Security Requirements

Before the 1990s, most aircraft systems were independent and self-contained, sometimes even physically isolated, creating air gaps. After this decade, the systems started to become more and more integrated, using digital systems to exchange information or performing software updates, removing the air gaps. From the cybersecurity point of view, the aircrafts started to be a potential target for attacks.

Currently, as aircraft are becoming like commercial Information Systems, using COTS systems such as Operational Systems, they are also exposed to the same security-threats of a ground equivalent system. Due to this fact, the aeronautical engineers shall consider not only "regular" safety requirements, but also the security requirements and their impact on the aircraft.

On other hand, the hackers were becoming more ambitious and the malwares sophisticated. A famous case which started a new era of sophisticated malwares is the Stuxnet. Stuxnet was developed specifically to attack the SCADA control system and it shows how the hackers may use the same strategy to attack any other complex and critical system.

10.2.1 Understanding the Stuxnet

Specifically developed to attack supervisory control and data acquisition systems, the Stuxnet was the first one created to attack a complex and critical industrial control system and started a new generation of malicious code towards real-world infrastructure.

For a long time, the attackers collected any possible information regarding the Natanz nuclear site, including PLC controllers used to control the centrifuge motors. After gathering all the information needed, the attackers developed a specific worm to be recognized as legitimate software by the operational system. After that, the operating system services were available to the worm, which took control of the system. Some centrifuges spun so violently that they simply broke and became unusable. The worm was so specific that it was capable of fake the generated reports indicating that everything was running fine.

The centrifuge network from Natanz nuclear site was segregated from the Internet via an air gap. So, despite spending some time, the infection happened as expected, using USB connection to upload the worm in the segregated network.

This infection case showed to the world that an attack may involve a lot of time and resources, and may be even more complex and sophisticated than the target itself.

10.2.2 May Stuxnet Similar Case Also Happen in Aircraft?

If we consider only safety threats during an aircraft system requirements development, we may be exposed to attacks. For example, many aircraft use field-loadable software to update software version without being necessary to move the aircraft system to a specialized laboratory. For safety-related system, the classical development requests a lot of protection mechanism which may be sufficient to prevent attacks. But if we consider that some aircraft are using a nonsafety system to introduce data in the aircraft network, like software loads and aeronautical databases; we have an open door for an attacker.

Protection of the aircraft from security attacks requires both physical and aircraft systems measures.

10.2.3 But are the authorities doing something in this new scenario?

The answer is yes, the authorities, together with the aeronautical industry, developed a set of airworthiness security processes, they intend to complement the existing standards for convectional development.

1. DO-326A/ED-202A: which presents the main guidelines and security considerations.

2. DO-356A/ED-203A: which is a "how-to" for the DO-326A/ED-202A.
3. DO-355A/ED-204A: which is a "in-service" guidance, such as Operation, Support, Maintenance, Administration, and Disposal.
4. ED-201, ER-013, ER-017: which are the support guidelines.
5. ED-205 for ground systems, it is mainly used in Europe.

EUROCAE and RTCA are planning to produce more security-related guidelines in the future, such as an updated version for ED-205, ER-013, and ED-201, and a RTCA document equivalent to ED-205A, ER-013, and ED-201A.

10.2.4 Understanding the DO-326A/ED-202A Airworthiness Security Process Specification

DO-326A/ED-202A [2, 5] is the core security standard published in 2010 and updated in 2014. Basically, it introduces processes described to complement the aircraft system development with security concerns. It defines security concepts such as Security Scope Definition, Security Risk Assessment, and considerations about the Security Development activities to be performed during the aircraft development process.

Interactively with ARP 4754 [8] activities, DO-326A/ED-202A [3, 6] introduces the security risk assessment process to system development and safety assessment processes. The security process will consider the security issues and provide security requirements to the System Development process.

It also provides guidance for modifications to aircraft and systems, including Supplemental Type Certification and Amendments to Type Certification.

Details of the Airworthiness Security Process activities are introduced in DO-326A/ED-202A [3, 6] Appendix A, it defines the input and outputs, compliance objectives and dependencies of the activities, including the interaction with safety assessment or system development processes, as per ARP 4754 [8].

The activities are organized in three categories:

1. **Certification related activities:** activities related to compliance demonstration with airworthiness regulation. The plan for security aspects of certification and the accomplishment communication summary are defined in this section.
2. **Security Risk Assessment related activities:** the aircraft and system security scope and risk assessment are defined in this section.

3. **Security Development related activities:** activities to evaluate the security effectiveness of the security measures and ensure that the identified risks are mitigated. The security requirements are determined in this category.

10.2.5 Why Do We Need Specific Guidelines for Security Requirements?

The DO-178/ED-12 standards for airborne software have been in use since 1981, at that moment, the guideline did not consider security aspects. In 2006, the EUROCAE created the Working Group 72 and, in 2007, the RTCA created the Special Committee 216 to develop the DO-326/ED-202 airworthiness security process specification.

Both standards were designed in different moments of the aeronautical industry history. DO-178/ED-12 was designed to guide the airborne software development considering safety impacts only, while the information security may impact safety or not, but other aspects should also be considered such as operational, commercial, intellectual property, and legal aspects. In addition, security is a unique discipline, which requires its own specialists. These specialists must have different expertise than the airworthiness safety specialists and need to perform specific security analysis, such as threat analyses.

Another point is conceptual, AC/AMC 25.1309 explicitly excludes "sabotage" from the list of events to be addressed during the safety assessment. Therefore, instead of updating the DO-178/ED-12 or even ARP-4754, the decision was to create separate standards to address security aspects.

10.2.6 A Practical Example of a Possible Back Door for an Attacker

Currently some aircraft systems are comprised of an integration of software components, which are uploaded to the LRU using field-loadable capabilities without removal of the hardware from the aircraft. Furthermore, some of these systems support Electronic Distribution Software to delivery software updates to be loaded in the units without using physical media. A variety of software can be transferred using the EDS systems such as navigation and maintenance databases.

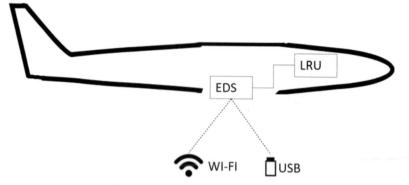

Figure 10.1   EDS system interfaces.

These EDS systems are usually classified as nonsafety critical by the safety assessment and may be developed with a lower level of rigor than the critical systems, avoiding a lot of activities requested by the classical aeronautical development. But if we consider that the EDS system has access to the Ethernet connection, interconnects with other LRUs for the transfer of field loadable software and these LRUs may be safety-critical, then we have a potential threat for the entire aircraft.

Figure 10.1 shows a typical EMS system used in some aircraft, it is connected to one or more LRUs via AFDX protocol and provides Wi-Fi and USB connection to receive data during the maintenance of the aircraft. When the aircraft is on ground the EMS system may receive updates using the Wi-Fi or USB interface and transmit them to the LRUs. At this moment, the EMS system is vulnerable to threats entering through its interfaces, attacks may originate in portable devices that are temporarily connected to the system, for example. Therefore, we need to determine the security perimeter considering the system's logical and physical interfaces with the aircraft and the outside world.

While systems inside the security perimeter are controlled by design. Analyzing Figure 10.2, it is clearly that the EMS system is a connection to an uncontrolled world and it may be a threat source. In the security scope, the LRU that interfaces with the EMS system shall also be considered as external even if it is embedded in the aircraft.

DO-356/ED-203 [3, 6] provides a set of survey questions which intend to help the identification of potential sources of vulnerability. Questions about the use of operating systems, network interfaces to share data, and removable

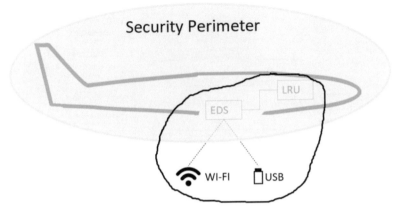

Figure 10.2 Defining the security perimeter.

media and electronic devices, like external hard disk or USB sticks, are very useful during the vulnerability analysis.

Usually, EMS systems use COTS operating system, which has a lot of third-party libraries. Considering the survey questions, it is right to say that this non-safety critical EMS system is an open door for an attacker and it should be considered as a security-critical system. During the requirements capture process of this EMS system, the security assurance shall be considered, and the severity of the threat condition effects on aircraft safety shall determine the development assurance actions for all functions and components of this system.

After the security scope is defined, the security risk assessment may be performed to identify the threat conditions and their impact on safety. In our EMS system, threats arise from devices connected to the Wi-Fi or USB interface. We will analyze this security threat in detail in the Section 10.2.8.

10.2.7 Considering Security Aspects During the Aircraft Development Lifecycle

Figure 10.3 summarizes how the process model for aircraft and system development is integrated with security activities, showing the relation of ARP-4754 [8] and DO-326/ED-202 [3, 6]. To facilitate comprehension, ARP-4754 [8] system development processes are depicted with white boxes, and the additional security processes required by DO-326/ED-202 [3, 6] are depicted as dashed gray boxes.

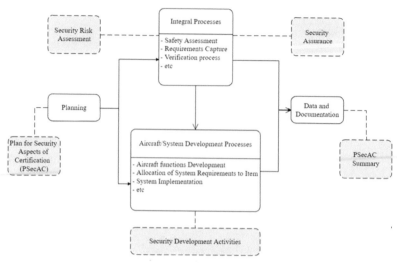

Figure 10.3   Security activities in the ARP-4754 aircraft development process.

During the planning phase, the PsecAC should be created containing the relationship or dependencies to other systems certification plans, including means to show compliance with security regulatory requirements and planned iterations with certification authorities.

The Security Risk Assessment uses outputs from safety assessment process and may provide updates to aircraft and system requirements. The Security Risk Assessment shall interact with Safety Assessment process to determine how security events impacts to aircraft safety. The risk assessment activities should be performed at aircraft and system levels.

Integrated in the ARP-4754 [8] Integral Processes, the Security Assurance process assures that the security measures perform as intended and their effectiveness against known and unacceptable vulnerabilities.

During the development process is possible to identify some items that require security involvement and determine the security requirements. These items such as busses, operating systems, software, and programmable hardware may introduce vulnerabilities that can be exploited to compromise the aircraft systems. In this phase, the security requirements are also validated and verified.

In the end, the PsecAC Summary will describe and justify any deviation from the PsecAC and provide evidence on activities performed to show compliance with the applicable objectives.

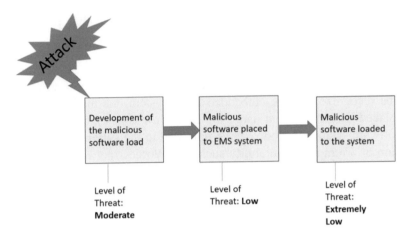

Figure 10.4   Security threat scenarios.

10.2.8 Defining Security Treat Conditions

Let us consider that, during the security risk assessment of the EMS system provided in Section 10.2.6, the following security threat was identified as having an impact on system safety: "T-010-Unlicensed data being loaded into the system".

After the threat conditions and their severity have been determined, security specialists shall analyze each threat scenario to determine their risk acceptability. As stated in DO-326/ED-202 [3, 6] the risk acceptability of a threat scenario is defined by:

- The level of threat of the scenario.
- Severity of effect of the threat condition caused by the threat scenario.

Threat condition T-010 identifies that malicious software updates may compromise the behavior of the system. It comprises three security threat scenarios, as depicted in Figure 10.4.

Scenario 1: Development of the malicious software load.

The entry point for an attacker is developing the malicious code to infect the system. The malicious code may be created using reverse engineering or even taking information from another aircraft that also uses the same system. The attacker should have access to the system software or another obsolete version of the software to gather the needed knowledge to create the infected software. Despite this software requires development with specialized knowledge, it may have been reused from another aircraft's development. The security level of this threat scenario is "Moderate".

Scenario 2: Malicious software placed to EMS system

If the attacker gathered sufficient information to create a piece of software that can be recognized by the embedded EMS system, the malicious code may be placed into the system using the same procedures required by the official software. Malicious code can be uploaded to the system through the exploitation of a remote vulnerability such as a Wireless network or HTTP server. Despite the embedded EMS system is a system that requires deep knowledge about the system characteristics, the vulnerabilities of its remote interfaces are well known by the attackers, the security level of this threat scenario is "Low."

Scenario 3: Malicious software loaded to the system.

Once an incorrect data load is created and placed in the system, it needs to be loaded into the system. Usually, data loading is only possible when the aircraft is on the ground and some protection mechanisms are activated. Therefore, only authorized staff may have physical access to the aircraft. The probability of an authorized user intentionally adds incorrect data to the system is reduced considerably. Besides that, to load the malicious software into the system it must have a valid CRC, the security level of this threat scenario is "Extremely Low."

Our threat condition analysis showed that we have an exploitable scenario that may be an entry point to an attacker. Despite the performed analysis provided a low level of threat, an attacker may use the exploitable vulnerabilities from remote interfaces using an infected malicious software load. To mitigate this security risk, security measures will be adopted, and security requirements will be developed and allocated to the system according to the system architecture during the security development activities.

10.2.9 Security Measures

Security Measure is used to protect the confidentiality, integrity, and availability of the system. Security measures may include onboard or external procedures. They may be technical using software or hardware solutions such as firewalls, or operational such as policy and practices.

According to DO-326/ED-202 [3, 6], security measures should be identified during the development of the security architecture and requirements created to implement the security measures.

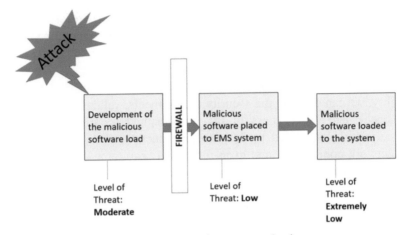

Figure 10.5   Security measure adoption.

In T-010, a Firewall may be adopted in the operating system for example. It would deny all unauthorized incoming and outgoing connections, allowing only an approved connection, and reducing the risk of an attack, as depicted in Figure 10.5.

10.2.10 Developing Security Requirements

Security requirements define new requirements or additions to existing system requirements to address security issues or mitigate exploitable vulnerabilities. Security requirements are derived from industry standards, applicable regulatory requirements, and a history of exploited vulnerabilities in the past. According to DO-356/ED-203 [4, 7], security requirements should be implemented in the design in a way so that safety and security are coordinated. DO-356/ED-203 [4, 7] also defines that security requirements may be analogous to derived requirements because, usually, there are no higher-level requirements or aircraft/system functions as a source to the security requirements. It is also important to note that security requirements may consider both safety and business needs.

During the requirements capture process of the system or item, regular and security derived requirements shall be analyzed to determine their impact on safety and security assessment.

One singular characteristic of the security requirements is that they may be handle different from the regular requirements. Since in some cases the security requirements must be avoidable, they may not be verifiable only by test, but a combination of test/analyses/demonstration. Therefore, some rules applicable to regular requirements may not be applicable to security requirements and it is recommended by ED/203A to clearly identify the security requirements to simplify dedicated tracing of them.

After analyzing each threat condition scenario to stablish the risk acceptability, we can develop the security requirements, validate, and verify them to check the effectiveness of the implemented security measures. Security requirements related to interfaces connection, security log, data loaders, network connection, and even functional requirements to security mitigations should be considered during the security development activities.

The Table 10.1 shows some examples of security requirements that will be implemented to mitigate the security risks of the threat condition T-010. It is important to note that the provided security requirements examples are illustrative only, they may not be complete comparing to real development needs.

To exercise the effectiveness of the designed security measures, the implemented security requirements must be verified. The Security tests are comprised of vulnerability analysis and penetration tests. The vulnerability analysis typically searches for any known vulnerability presented in the system without exercising them to avoid damages; the penetration tests are executed to exploit the effectiveness of the implemented security measures revealing the system behavior against malicious attacks.

Usually, vulnerability analysis is done using a COTS security tool in an automatic way. On other hand, penetration tests are manually executed and should be traceable to security requirements like regular DO-178/ED-12 [2, 5] or ARP-4754 [8] tests. The penetration tests report shall contain, at least, information about the test environment, description of the test, test steps to be performed, pass/fail criteria and security requirement related to each penetration test. Table 10.2 adds to security requirements some test scenarios to verify them. It is important to note that the provided test scenarios examples are illustrative only, they may not be complete comparing to real development needs.

During the tests execution it may be identified a deviation from the requirements, showing that some security requirements were not properly implemented. In this case, the implanted security measures should be reassessed to determine the impact of the deviations in the threat conditions.

Table 10.1 Security requirements.

ID	Description
001	The system wireless device shall connect to the configured access points using WEP 128-bit security protocol.
002	The system USB interface shall transfer data only with devices that are USB MSC compliant.
003	The system shall create a history log containing the following information: • Time • Date • User • Operation performed
004	The system shall write down, in an history log, successes and failures attempts of authentication to access the system.
005	The history log shall be protected from modifications.
006	The USB interface shall be deactivated when the aircraft is on the flight.
007	A firewall mechanism shall be configured in the system wireless interface. *
008	The system shall permit only software applications presented in the whitelist to be installed to the operation system. *
009	The system shall permit loading of an application placed into to the system only if the aircraft is on ground and the physical mechanism is enabled.
010	All the authentication access to the system shall be comprised by a username and password.

*This requirement considers that the system is using a CTOS operational system, such as Windows Embedded Standard.

10.3 Conclusion

Historically security and safety were segregated technologies with separate fields of concerns. While security was only considered in system which have an external connection to the world, such as information management; safety was the only concerned to aircraft developers. However, the aircraft have become like regular integrated systems and are connected to the external world through attackers' well-known technologies.

Cybersecurity should be integrated into the aircraft processes through processes at aircraft development level, adding further security requirements and activities to the regular development. But the security standards also require additional security specific activities and objectives to reduce to an acceptable level, the threats that may impact on the aircraft safety.

Table 10.2   Test scenarios for the security requirements.

Req ID	Requirements Description	Test Scenario
001	The system wireless device shall connect to the configured access points using WEP 128-bit security protocol.	Scenario 1: Verify that the system can connect to access point using WEP 128-bit security protocol. Scenario 2: Verify that the system cannot connect to access point using another security protocol.
002	The system USB interface shall transfer data only with devices that are USB MSC compliant.	Scenario 1: Verify that the USB interface interfaces with devices that are compliant with the USB MSC. Scenario 2: Verify that the USB interface does not connect with devices that are not compliant with the USB MSC.
003	The system shall create a history log containing the following information: • Time • Date • User • Operation performed • Operation status (success/failure)	Scenario 1: Analyze the history log to check if only the required information was recorded.
004	The system shall write down, in an history log, successes and failures attempts of authentication to access the system.	Scenario 1: Analyze the history log to check if all authentication attempts were recorded, including the unsuccessful ones.
005	The history log shall be protected from modifications.	Scenario 1: Attempts to modify the recorded information in the history log shall be performed to check if it is protected against modifications.
006	The USB interface shall be deactivated when the aircraft is on the flight.	Scenario 1: Verify that USB interface is disabled when on ground discrete is disabled.
007	A firewall mechanism shall be configured in the system wireless interface. *	Scenario 1: Analyze that the firewall is properly installed and configured in the system. Scenario 2: Verify that the firewall is properly working as configured.

Continued

Table 10.2 *Continued*

Req ID	Requirements Description	Test Scenario
008	The system shall permit only software applications presented in the whitelist to be installed to the operation system. *	Scenarios 1: Verify that all applications presented in the while list were properly installed in the system. Scenarios 2: Verify that an application outside the whitelist was not installed in the system.
009	The system shall permit loading of an application placed into to the system only if the aircraft is on ground and the physical mechanism is enabled.	Scenario 1: Verify that the system allows application load according to table below:

On ground discrete	physical discrete	Load
true	false	false
false	true	false
false	false	false
true	true	true

Req ID	Requirements Description	Test Scenario
010	All the authentication access to the system shall be comprised by a username and password.	Scenario 1: verify that the system allows access only to a valid username and password. Scenario 2: verify that the system denies access to an invalid username and/or password, including blank values.

We should also keep in mind that attackers may not be only motivated for money, but also for religion and politician aspects, or they only want to have their Warhol moment. Nevertheless, to be well prepared for zero-day attacks, we need to consider the security requirements in the same level of importance that the safety requirements during the aircraft development process.

References

[1] Symantec,W32.Stuxnet Dossier, 2010.
[2] Radio Technical Commission for Aeronautics, DO-178C Software Considerations in Airborne Systems and Equipment Certification, Washington, D.C., 2011.
[3] Radio Technical Commission for Aeronautics, DO-326A Airworthiness Security Process Specification, Washington, D.C., 2014.

[4] Radio Technical Commission for Aeronautics, DO-356A Airworthiness Security Methods and Considerations, Washington, D.C., 2018.
[5] European Organisation for Civil Aviation Equipment, ED-12C Software Considerations in Airborne Systems and Equipment Certification, France, 2012.
[6] European Organisation for Civil Aviation Equipment, ED-202A Airworthiness Security Process Specification, France, 2014.
[7] European Organisation for Civil Aviation Equipment, ED-203A Airworthiness Security Methods and Considerations, France, 2018.
[8] SAE Aerospace, ARP-4754A Guidelines for Development of Civil Aircraft and Systems, France, 2018.

11

Requirements Engineering Maturity Model for Safety-Critical Systems

Jéssyka Vilela

Universidade Federal de Pernambuco, Brazil
E-mail: jffv@cin.ufpe.br

Abstract

Assessing the maturity level of software development companies plays an important part in implementing a systematic and well-directed approach for process improvements. Such assessments provided by maturity models contribute to obtaining a certain quality, reducing requirements issues and errors, and verifying companies' capabilities on a comparable basis. In the development of a Safety-Critical System (SCS), stakeholders should deal with many suppliers, handle several features, for instance, organizational, technical, strategic) and this requires specialized processes, techniques, skills, and experience. Consequently, new challenges emerged to be handled by Requirements Engineering (RE). This chapter describes the concepts and fundamentals of Uni-REPM SCS, a safety module for the Unified Requirements Engineering Process Maturity Model (Uni-REPM). It consists of seven main processes, 14 subprocesses, and 148 safety actions that define principles and practices that constitute the basis of safety processes maturity. Uni-REPM SCS can support companies in assessing their current practices and offer a stepwise improvement strategy to achieve higher maturity. Finally, we conclude this chapter with a real example of safety maturity evaluation.

Keywords: Safety-Critical Systems, Maturity Models, Uni-REPM, Software Quality.

11.1 Introduction

The software is currently the primary source of risks and hazards of a Safety-Critical System (SCS) due to its ability to control the hardware through actuators [2, 13, 17]. If the software provides wrong instructions, accidents, and safety incidents can happen with significant consequences, including loss of lives and substantial financial losses [1, 3].

The severity and impact of such accidents demand that safety concerns shall be handled early in the system development process [13, 16]. Accordingly, there is a need for incorporating correctness into software as it is being developed. This approach, called cleanroom software engineering [22], has the purpose of avoiding dependence on costly defect removal processes by writing functionalities right from the beginning and verifying their correctness before testing.

However, to properly follow such development approach, stakeholders should handle several features (e.g., organizational, technical, and strategic) [18], and this requires specialized processes, techniques, skills, and experience [19, 20]. Accordingly, engineers must plan and specify SCS carefully, requiring sophisticated software engineering approaches [2, 3, 13] to build safety-critical software with high quality [21].

Software quality is an area of software engineering, whose goal is to promote an effective software process applied in a manner to create a useful product that provides measurable value to those who produce and those who use it [22]. In an SCS, the most crucial quality dimension is *reliability* which is concerned with developing a fault-tolerant system, that is, the system is available when requested, and it provides resources and capabilities without failures [22].

Software quality addresses different aspects [3, 22]:

- *A useful product:* the quality management team must evaluate whether a system meets its specification [3]. Satisfaction is a subjective aspect; hence, some judgment should be performed along with meetings with the client and test results to decide if an acceptable level of the system's characteristics has been achieved. This evaluation involves analyzing the programming and documentation standards followed, the tests conducted, and the software structure and understandability.
- *Non-Functional Requirements (NFR) satisfaction:* software quality is not just about whether the software functionality has been correctly implemented but also depends on its NFRs. These requirements

correspond to constraints on the services or functions offered by the system, such as safety, performance constraints, reliability, store occupancy, security, or availability.

- *Aggregating value for both the developer and user:* high-quality software provides benefits for the software organization like less maintenance effort, fewer bug fixes, reduced customer support, and the end-user community such as accelerating some business process.

- *Effective software process:* proper processes are more likely to lead to software with higher quality and lead to fewer defects in the system. Accordingly, beneficial software development practices should be used to improve the quality of the developed software.

In this chapter, we address the quality of safety-critical systems from the point of view of software process improvement (SPI). According to Pressman [22], this means that we consider a set of activities that will lead to an improved software process and higher-quality software delivered in a more timely manner. In SPI initiatives, we should evaluate the software development process using an SPI framework or maturity model [22].

Maturity models rely on a set of processes and actions describing software engineering practices or actions of effective processes at distinct phases of development that a software company must follow to improve its software development process [5].

A software process consists of a set of activities conducted in the production of a software product. There are several software processes proposed in the literature, but all usually comprise four activities that are central to software engineering [3]:

- *Software specification:* It corresponds to determining the functional and nonfunctional software requirements.
- *Software design and implementation:* It involves the definition of software architecture, low-level requirements, and software coding.
- *Software validation*: It consists of conducting tests, reviews, inspection in order to ensure that the software fits its intended purpose.
- *Software evolution:* It comprises the activities related to software maintenance and evolution.

The focus of this chapter is the *maturity-based process improvement* regarding the *safety* nonfunctional requirement. Accordingly, the companies should assess their processes' maturity to indicate their safety processes quality. This evaluation allows them to understand their existing safety processes and change them to increase system quality, reduce costs and development time, contributing to the development of better software. This approach is vital to large business companies, safety-critical systems, and systems involving developers in different companies in which management issues are often the reasons why projects run into problems [3].

The software specification is the software development phase in which the functional and non-functional software requirements are elicited and documented. This phase is commonly known as Requirements Engineering (RE). It corresponds to the first stage of a software development process that addresses all of the activities involved in discovering, documenting, and managing a set of requirements for a system [12].

A structured set of activities compose the RE process that includes [12]: Requirements elicitation, Requirements analysis and negotiation, Requirements documentation, Requirements validation, and Requirements management. The RE phase is fundamental in the software development process. RE issues such as vague initial requirements, ambiguities in the requirements specification, undefined requirements process, requirements growth, requirements traceability, and confusion among methods and tools have a significant impact on the quality of a system, especially in a safety-critical one. In this context, it is claimed that the most cost-efficient place to fix many of the mentioned problems is in the RE phase.

Considering the benefits of improving the RE process, the Uni-REPM maturity model was proposed by Svahnberg et al. [14] to assess the strengths and weaknesses of companies' software processes. In this chapter, we describe a safety module for the RE process of safety-critical systems.

11.2 A Maturity Model for Safety-Critical Systems

Uni-REPM SCS is a safety module proposed for the universal lightweight Uni-REPM maturity model [14]. Aiming to maintain compatibility with Uni-REPM, the safety module follows its dual-view approach, and the existing process artifacts were not altered, and none were removed.

The safety module was proposed to reduce issues in RE during SCS development by addressing safety actions/practices that should be covered in the RE process [2] to reduce the gap among RE and safety engineering [15, 16]. Uni-REPM SCS is structured in two views: Process Area and Maturity Level described in the following sections.

11.2.1 Process Area View

The process area view is used to visualize the practices that logically belong together in a fast way. This process area view specifies the model hierarchy in three degrees of depth:

1. *Main Process Area (MPA)*: It is an element that corresponds to the main activities of requirements engineering.
2. *Subprocess Area (SPA)*: It is an element that group actions related to a specific area that, when implemented correctly, contribute to achieving the goals considered important for improvement in this area.
3. *Actions*: safety practices considered important to the SPA, to which it is associated, be achieved.

On the top level of the model, there are seven MPAs corresponding to RE main activities (Requirements elicitation, Requirements analysis and negotiation, Requirements documentation, Requirements validation, and Requirements management [12]) and two more categories for organizational support and release planning that was not explicitly covered previously [14]:

1. *Organizational Support (OS)*: evaluates the quantity of support provided to RE practices from the surrounding organizations.
2. *Requirements Process Management (PM)*: comprehends activities to manage, control requirements change, and assure that the process is being followed.
3. *Requirements Elicitation (RE)*: it contains actions for discovering and understanding customers' necessities and desires to communicate them to other stakeholders.
4. *Requirements Analysis (RA)*: it comprises activities to detect errors, create a comprehensive view of requirements, and approve information needed in later activities of the RE process.
5. *Release Planning (RP):* comprises necessary actions to define the optimal set of requirements for a specific release aiming to accomplish defined/estimated time and cost goals.
6. *Documentation and Requirements Specification (DS):* it deals with how a company organizes the requirements and other information collected throughout elicitation into consistent, accessible, and reviewable documents.
7. *Requirements validation (RV):* it embraces checking the requirements against defined quality standards and the real needs of several stakeholders. It aims to ensure that the documented requirements are complete, correct, consistent, and unambiguous.

Each MPA is further broken down into fourteen SPAs [14], which contributes to a better organization of related practices and the ability of a company to plan its organizational efforts. At the bottom level, an Action denotes a specific activity that should be done or a particular item that should be present. The safety model extends Uni-REPM by adding 14 new subprocess areas, as listed in Table 11.1.

Figure 11.1 presents the Safety module and its relationship with Uni-REPM. The module extends the Uni-REPM model by adding 14 new SPAs highlighted through dashed lines.

In the safety module, actions also follow the same format adopted by Uni-REPM. Actions are identified by the MPA/SPA under which they belong, followed by an "a" which stands for "action" and their position in the group. For example, action "RA.PSA.a5 Specify the type of initiating events that need to be considered" of Figure 11.2 means the first action under MPA "Requirements Analysis" and SPA "Preliminary Safety Analysis."

Besides the description of each action, it can have *Supporting actions* that are the list of actions related to a specific action reflecting the high-level relationship between them as well as *Examples* that give practitioners suggestions on proven techniques or supporting tools when performing the action.

11.2.2 Maturity Level View

The Maturity Level establishes a level to each action (from 1 to 3), corresponding to the "Basic," "Intermediate," or "Advanced" level, depending on the difficulty to implement the action, how essential it is for the RE process, and dependencies among actions.

According to [3], a process maturity level indicates how good technical and management practices have been implemented in the company software development processes. Process improvement is a long-run project in which each step in the improvement process possibly will last quite a lot of months. For this reason, it is a continuous activity considering that when new processes are introduced, the company will be transformed, and the new processes will themselves have to evolve to accommodate such transformations.

The complete description of 148 actions of Uni-REPM SCS following the structure of Figure 11.2 can be found on the project website.[1]

---

[1] www.unirepm.com

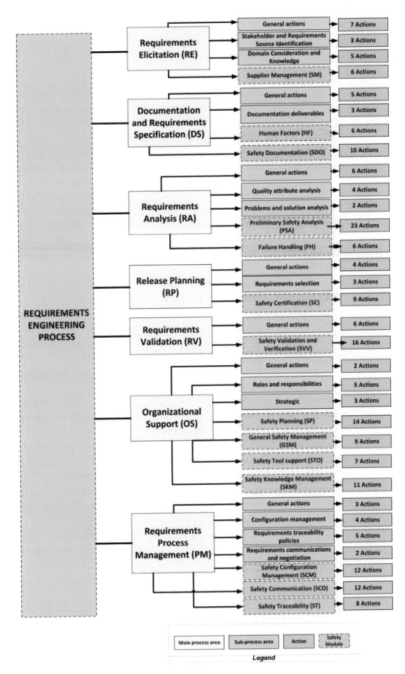

Figure 11.1 Safety module and its relationship with Uni-REPM [2].

Table 11.1 Overview of Uni-REPM SCS safety subprocesses.

UNI-REPM MPA	New safety subprocess area	Description
OS	Safety Knowledge Management (SKM)	It gives transparency in the development process by ensuring that projects and the company have the required knowledge and skills to accomplish the project and organizational objectives.
	Safety Tool support (STO)	It is responsible for facilitating the appropriate execution of the corresponding tasks and manage all safety-related information that should be created, recorded, and adequately visualized.
	General Safety Management (GSM)	It comprehends project safety management activities related to planning, monitoring, and controlling the project.
	Safety Planning (SP)	It provides safety practices and establishes a safety culture in the company.
PM	Safety Configuration Management (SCM)	It handles the control of content, versions, changes, distribution of safety data, proper management of system artifacts, and information relevant to the organization at several levels of granularity.
	Safety Communication (SCO)	It aims to improve the safety communication subprocess by defining actions related to safety terms, methods, processes to assist the safety analysis and assurance processes.
	Safety Traceability (ST)	It covers the traceability among artifacts, helping determine that the changes' requirements have been thoroughly addressed.
RE	Supplier Management (SM)	It is responsible for managing the acquisition of products and services from suppliers external to the project, which shall have a formal agreement.
RA	Preliminary Safety Analysis (PSA)	It addresses the realization of a preliminary safety analysis to avoid wasting effort in the next phases of system development.
	Failure Handling (FH)	It handles failures in system components that can lead to hazardous situations, the addition of redundancy, and protection mechanisms.
RP	Safety Certification (SC)	It has actions related to system certification.
DS	Human Factors (HF)	It handles issues regarding the system's users and operators that can lead to hazards and shall be handled during the RE phase of safety-critical system development.

Continued

Table 11.1 Continued

UNI-REPM MPA	New safety subprocess area	Description
	Safety Documentation (SDO)	It contains practices to record all information related to the system's safety produced in the RE phase.
RV	Safety Validation and Verification (SVV)	It contains actions to requirements validation and the definition of strategies to the verification of requirements aiming to obtain requirements clearly understood and agreed by the relevant stakeholders.

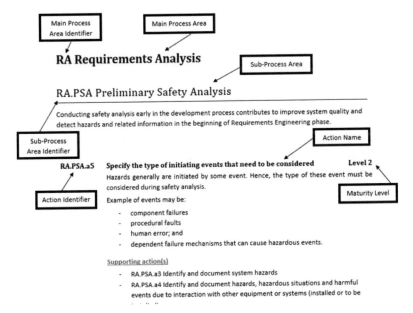

Figure 11.2 Example of Uni-REPM safety module action [2].

11.3 Evaluating the safety processes

11.3.1 Assessment Instrument and Tool

The Uni-REPM safety module follows the same approach of Uni-REPM by developing an assessment instrument in which the evaluator should select one of three options:

- "Incomplete" (vital action performed partially or not at all in the RE process);

- "Complete" (action was completed in the RE process); and
- "Inapplicable" (action was not necessary or possible to be performed in the process).

The evaluation can be performed by stakeholders involved in the RE process of an SCS, for instance, Requirements Engineer, Safety Engineer, Software Engineer, and Quality assurance engineer.

Evaluators often find maturity assessments a tedious activity since many maturity models do not have software tool support requiring manual work or the use of spreadsheets. If tool support is available [3], minimal effort must follow and evaluate the company's maturity.

The assessment instrument in Uni-REPM SCS is implemented in a web software tool aiming to facilitate and automate the maturity evaluation process. The tool is available at <www.unirepm.com>. The tool supports the evaluation of both Requirements and Safety processes supporting three types of user:

(1) external evaluator – this user can insert companies, projects and perform Safety/RE evaluations;

(2) internal evaluator – this user only can insert projects to the company he/she belongs and perform Safety/RE evaluations; and

(3) admin – besides the functionalities of external evaluators, this user can manage users and different versions of RE/Safety models.

The maturity level is determined when all questions are answered considering: (1) for each SPA, all actions at a certain level must be Completed (or Inapplicable) to the MPA achieve such level and (2) for the whole process, all actions at a certain level must be Completed (or Inapplicable) to the process achieve such level.

11.3.2 Results of a Safety Maturity Assessment

In this section, we debate the results of a maturity level evaluation considering the structure of Uni-REPM SCS in an actual project. The results showed that the project skips essential steps during RE phase; henceforth, it obtained level *Incipient*, that is, it did not satisfy all the actions of level 1 (Basic).

Figure 11.3 shows the total number of safety actions performed by the evaluated project in each of the seven MPAs. In the X-axis are arranged MPAs, the Y-axis is the number of safety practices performed by MPA, and colors represent the number of complete practices by the project according to the chart's legend.

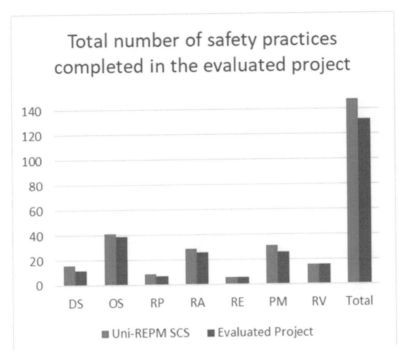

Figure 11.3 Comparison of the total number of actions of all MPAs of the evaluated project.

Table 11.2 shows an overview of the assessment results regarding the seven MPAs of Uni-REPM SCS. We can observe that although the evaluated project was classified as level *Incipient*, some MPAs were classified as Intermediate or even as Advanced level. This classification as *Incipient* happens since some basic safety actions were not satisfied by the project. Thus, it did not achieve the Basic level for the entire project. These results also mean that the company uses all parts of the Uni-REPM SCS.

The same comparison regarding the SPAs is presented in Figure 11.4. It shows the total number of satisfied safety actions in SPA by the project. In the *X*-axis are arranged SPAs, the *Y*-axis is the number of safety practices performed by SPA, and colors represent the number of complete practices by the project according to the chart's legend.

From Figure 11.4, we conclude that the actions of six SPAs: Human Factors (HF), General Safety Management (GSM), Safety Tool support (STO), Failure Handling (FH), Supplier Management (SM), and Safety Validation and Verification (SVV) were completely satisfied by the evaluated project.

Table 11.2   Overview of Uni-REPM SCS safety subprocesses.

MPA	Uni-REPM SCS	Evaluated Project	MPA Level achieved
Documentation and Requirements Specification	16	12	Advanced
Organizational Support	41	39	Zero
Release Planning	9	7	Intermediate
Requirements Analysis	29	26	Advanced
Requirements Elicitation	6	6	Advanced
Requirements Process Management	31	26	Zero
Requirements Validation	16	16	Zero
Total number of actions	148	132	

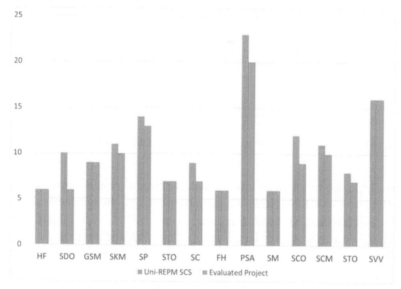

Figure 11.4   Comparison of the total number of actions of all SPAs of the evaluated project.

Table 11.3 shows the total number of safety practices at each maturity level proposed by Uni-REPM SCS and the percentage of complete actions per SPA and level.

In order to verify how close or distant the evaluated project is to reach a level of maturity, a latency analysis by maturity level can be performed. Figure 11.5 presents a latency chart of the evaluated project.

Table 11.3   Overview of Uni-REPM SCS safety subprocesses.

SPA	Uni-REPM SCS	Evaluated Project	% of complete actions	SPA Level achieved
HF	6	6	100	Intermediate
SDO	10	6	60	Advanced
GSM	9	9	100	Basic
SKM	11	10	90.91	Advanced
SP	14	13	92.86	Advanced
STO	7	7	100.00	Advanced
SC	9	7	77.78	Intermediate
FH	6	6	100.00	Intermediate
PSA	23	20	86.96	Advanced
SM	6	6	100.00	Advanced
SCO	12	9	75.00	Zero
SCM	11	10	90.91	Advanced
STO	8	7	87.50	Basic
SVV	16	16	100.00	Zero

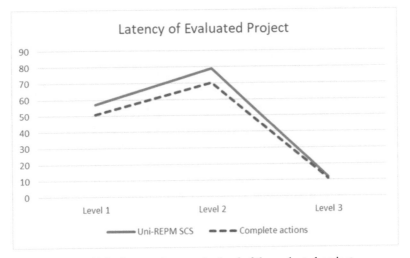

Figure 11.5   Latency by maturity level of the evaluated project.

This investigation will examine how many safety actions will be necessary for the maturity at each of the Uni-REPM SCS levels. The latency chart of Figure 11.5 shows that out of 57 proposed actions at the basic level, the evaluated project carries out 51 actions; at the intermediate level, from the 79 actions proposed by Uni-REPM SCS, the evaluated project performs 70; and in the advanced level, out of the 12 actions, it performs 11.

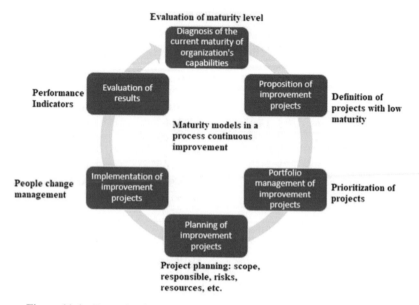

Figure 11.6   Example of maturity model usage for continuous improvement.

We illustrate the usage of safety maturity assessment results for continuous process improvement [22] in a company. We can define the process of process improvement as a cyclical process [3] inspired in the PDCA (Plan-Do-Check-Act) cycle proposed in Deming's lecture in Japan in 1950 [4]. First, the company should evaluate its capabilities, define the projects with low maturity, prioritize them, and plan the improvement (scope, responsibility, risks, resources, and other variables). Then, it should manage changes in people and company culture by implementing the improvement project, evaluate the results, and proceed with continuous improvement.

11.4 Conclusions

The maturity evaluation of software development processes contributes to obtaining better processes, a certain quality, reducing requirements issues and errors, and verifying companies' capabilities on a comparable basis. The evaluation of maturity of safety processes is necessary considering the specialized processes, techniques, skills, and experience involved in developing safety-critical systems.

In this chapter, we describe the concepts and fundamentals of Uni-REPM SCS, a safety module for Unified Requirements Engineering Process Maturity Model (Uni-REPM), along with a real example of safety maturity evaluation. The module consists of seven main processes, 14 subprocesses and 148 safety actions proposed to evaluate the safety processes maturity. The safety module can support companies in assessing their current practices as well as offers a stepwise improvement strategy to achieve higher maturity.

Acknowledgments

We would like to acknowledge that the KKS foundation partially supported this work through the S.E.R.T. Research Profile project[2] at the Software Engineering Research Lab, Blekinge Institute of Technology. We also want to thank the participants of the safety module validation for their availability to contribute to our research.

References

[1] N. G. Leveson. Safeware: system safety and computers, ACM, 1995.

[2] J. Vilela, J. Castro, L. E. G. Martins, T. Gorschek. Safety Practices in Requirements Engineering: The Uni-REPM Safety Module. IEEE Transactions on Software Engineering, 2018.

[3] I. Sommerville. Software engineering 9th Edition, 2011.

[4] W. E. Deming. Elementary Principles of the Statistical Control of Quality, JUSE, 1950.

[5] R. Wendler. The maturity of maturity model research: A systematic mapping study. Information and software technology, Elsevier, v. 54, n. 12, p. 1317–1339, 2012.

[6] P. Williams. A practical application of cmm to medical security capability. Information Management & Computer Security, Emerald Group Publishing Limited, v. 16, n. 1, p. 58–73, 2008.

[7] M. Johansson and R. Nevalainen. Additional requirements for process assessment in safety-critical software and systems domain. Journal of Software: Evolution and Process, v. 24, n. 5, p. 501–510, 2012.

[8] P. Fraser, J. Moultrie, M. Gregory. The use of maturity models/grids as a tool in assessing product development capability. In: IEEE International Engineering Management Conference, v. 1, p. 244–249, 2002.

---

[2] https://rethought.se

[9] T. D. Bruin, R. Freeze, U. Kaulkarni, M. Rosemann. Understanding the main phases of developing a maturity assessment model. Australasian Chapter of the Association for Information Systems, 2005.

[10] F. Marx, F. Wortmann, J. H. Mayer. A maturity model for management control systems. Business & information systems engineering, Springer, v. 4, n. 4, p. 193-207, 2012.

[11] T. L. Reis, M. A. S. Mathias, O. J. de. Oliveira. Maturity models: identifying the state-of-the-art and the scientific gaps from a bibliometric study. Scientometrics, Springer, p. 1–30, 2016.

[12] G. Kotonya and I. Sommerville. Requirements engineering: processes and techniques. Wiley Publishing, 1998.

[13] N. Leveson. Engineering a safer world: Systems thinking applied to safety. Mit Press, 2011.

[14] M. Svahnberg, T. Gorschek, T. T. L. Nguyen, M. Nguyen. Uni-repm: a framework for requirements engineering process assessment. Requirements Engineering, Springer, v. 20, n. 1, p. 91–118, 2015.

[15] J. Vilela, J. Castro, L. E. G. Martins, T. Gorschek. Integration between requirements engineering and safety analysis: A systematic literature review. Journal of Systems and Software, Elsevier, v. 125, p. 68–92, 2017.

[16] L. E. G. Martins, T. Gorschek. Requirements engineering for safety-critical systems: A systematic literature review. Information and Software Technology, Elsevier, v. 75, p. 71–89, 2016.

[17] P. Panaroni, G. Sartori, F. FABBRINI, M. FUSANI, G. LAMI. Safety in automotive software: an overview of current practices. In: Annual IEEE International Computer Software and Applications Conference, 2008. p. 1053–1058.

[18] B. Solemon, S. Sahibuddin, A. Ghani. Requirements engineering problems in 63 software companies in malaysia. In: International Symposium on Information Technology, 2008, v. 4, pp. 1–6.

[19] ISO. ISO/IEC TS 15504-10:2011 - Information technology - Process assessment - Part 10: Safety extension, 2011.

[20] SEI, Software Engineering Institute +SAFE: A Safety Extension to CMMI-DEV, version 1.2, 2007.

[21] P. J. Graydon, C. M. Holloway. Planning the unplanned experiment: Assessing the efficacy of standards for safety critical software. 2015. Available at: <https://ntrs.nasa.gov/search.jsp?R=20150018918>.

[22] R. S. Pressman. Engenharia de Software. 7th edition. Amgh Editora, 2009.

Index

About Editors and Authors

Aldo Martinazzo is pursuing Master's in Biomedical Engineering at Federal University of São Paulo (UNIFESP) in São José dos Campos, Brazil, with focus on methodologies for risk management of medical devices and their application on the project of a Low-Cost Insulin Infusion Pump. He holds bachelor's degree in Mechanical Engineering from Technological Institute of Aeronautics (ITA) in São José dos Campos, Brazil. Martinazzo has extensive experience in the development and safety assessment of safety-critical systems in the aeronautic industry.

Camilo Almendra is currently a Ph.D. candidate in Centro de Informática at Universidade Federal de Pernambuco, Brazil. His research interests include requirements engineering, safety–critical systems, and safety assurance cases. Currently, he is an Adjunct Professor at Campus Quixadá of the Universidade Federal do Ceará, Brazil.

Carla Silva is an Associate Professor at the Informatics Center of the Federal University of Pernambuco, Brazil. She had received the MSc and PhD degrees in Software Engineering from Federal University of Pernambuco (2001–2007). Her research interests are in Software Engineering, mainly in the following subjects: Requirements Engineering, Agile Methods and Adaptive Systems. She has been serving as a Program Committee member of conferences, such as RE, ICSE, ACM SAC RE Track, CibSE, CBSoft, and WER. Occasionally, she has served as reviewer for scientific journals, such as *Empirical Software Engineering*, *The Journal of Systems and Software*, *Journal of Software Engineering Research and Development*, *CLEI Electronic Journal*, and *Information and Software Technology*. She is also a supervisor of Ph.D. students.

Carlos Lahoz is a university professor at Sao Jose dos Campos, Brazil in Computer Science, emphasis in software engineering. Also, he is a collaborator professor at post-graduation course in Sciences and Space

Technology CTE at Technological Institute of Aeronautics, ITA, Brazil. His post-doctorate was in system safety at Aeronautics and Astronautics Department in Massachusetts Institute of Technology, MIT, USA. He received his PhD in system safety from Sao Paulo University USP/Brazil and Master thesis in space software quality at National Institute of Space Research INPE, Brazil. He was Senior Technologist for more than 30 years at Institute of Aeronautics and Space, IAE, Brazil and head of software engineer team of the Brazilian Space Launcher VLS-1.

Cristiano Constantino dos Santos has received his Bachelor's degree in computer science, and has 20 years of experience in software development using different methodologies. Cristiano is specialized in Design Quality Assurance in the Aeronautical Industry, which he has been working for more than 12 years, being responsible for assuring the development compliance with the aeronautical standards including software security standards. He is currently living and working in Germany, and is inspired daily by his wife and their two children. In his free time, Cristiano likes to watch movies, drink a good glass of wine, and play video games with his kids.

Jemison dos Santos is a technologist in Analysis and Systems Development at Federal Institute of Education, Science and Technology of Mato Grosso do Sul (IFMS). He received his Master's in Computer Science from Federal University of São Paulo (UNIFESP). Currently, he is a Lecturer at IFMS—Nova Andradina campus and Instructor of Higher Education at State University of Mato Grosso do Sul (UEMS)—Nova Andradina. He has experience in the areas of Software Engineering; Requirements Engineering; Relational Database; Software Testing, Front-end, Back-end Web Development and Research Methodology.

Jéssyka Vilela is Adjunct Professor at Universidade Federal de Pernambuco (UFPE). Previously, she was an Assistant Professor at Campus Quixadá of Universidade Federal do Ceará (UFC; 2017–2019). She has received PhD (2018) and MSc (2015) in Computer Science from UFPE. Graduation in Computer Engineering from Universidade Federal do Vale do São Francisco—UNIVASF (2012). Her main research lines include Safety-Critical Systems, Information Security, Software Engineering, Requirements Engineering, and Privacy with papers published in relevant journals and conferences. Her research has received best papers awards. She participates in extension projects as well as research projects with partnerships with researchers from national and international universities as well as innovation projects with public agencies.

Johnny Cardoso Marques is an Adjunct Professor in the Computer Science Division of Aeronautics Institute of Technology (ITA) in Brazil. His research interests are Software Quality, Safety-Critical Systems, and Requirements Engineering. He holds a degree in Computer Engineering from the Rio de Janeiro State University (UERJ), a Master's degree in Aeronautical Engineering and a doctorate in Electronic Engineering and Computing, these two from the Aeronautics Institute of Technology (ITA) in Brazil. He has over 20 years of experience in software engineering for the aviation industry. Currently, he contributes to several working groups at the IEEE Standards Association. Additionally, he has many international publications about Safety-Critical Software Development and acts as a reviewer for several international journals and conferences.

Sarasuaty Megume Hayashi Yelisetty received her degree in Computer Engineering from Vale do Paraiba University (UNIVAP) and received her Master's degree in Computer Engineering from Aeronautics Institute of Technology (ITA). She is currently a doctorate student at ITA. Additionally, she has been working at EMBRAER in software/airborne electronic hardware processes from past 10 years and has recognized experience in standards used for airborne system and software such as DO-178B/C, DO-254 and MIL-STD-498.

Lilian Michele da Silva Barros has 18 years of experience in aviation industry (EMBRAER) with expertise in certification standards for airborne systems and software development process. She started her career as an intern for 3 years at National Institute of Space Research (INPE) in Laboratory of Integration and Tests (LIT). She received a degree in Computer Science from University of Vale do Paraiba (UNIVAP) in 2003 and has been working as a Technician in Data Processing at the same institution from 1999. Currently, she is a Master's degree student at Aeronautics Institute of Technology (ITA) in Computer and Electronic Engineering.

Luiz Eduardo G. Martins has a Ph.D. in Electrical Engineering from State University of Campinas (UNICAMP), Brazil. Dr. Martins is an Associate Professor of software engineering and embedded systems at Federal University of São Paulo, where he works as a research leader in

204 About Editors and Authors

collaboration with industrial partners in safety-critical systems domain. His research interests include requirements engineering, software quality process, model-driven software development, IoT, and technological innovation in medical systems. Contact him at legmartins@unifesp.br

Tatiana Cunha is an Associate Professor at Institute of Science and Technology—Federal University of São Paulo (UNIFESP) in the area of Biomedical Engineering. Cunha received Master's degree and Ph.D. in Physiology. Cunha's research interests include the development and evaluation of a low-cost insulin infusion pump and therapeutic tools for the treatment of diabetes and emotional stress.

Tony Gorschek is a Professor of Software Engineering at Blekinge Institute of Technology, where he works as a research leader and scientist in close collaboration with industrial partners. Dr. Gorschek has over 15 years of industrial experience as a CTO, senior executive consultant, and engineer. At present, he works as a research leader and for several research projects developing scalable, efficient, and effective solutions when it comes to software-intensive product and service development. Dr. Gorschek leads the SERT profile (Software Engineering ReThought)—Sweden's largest software engineering research initiative.